OUR TRANSGENIC FUTURE

Our Transgenic Future

*Spider Goats, Genetic Modification,
and the Will to Change Nature*

Lisa Jean Moore

NEW YORK UNIVERSITY PRESS
New York

NEW YORK UNIVERSITY PRESS
New York
www.nyupress.org

© 2022 by New York University
All rights reserved

References to internet websites (URLs) were accurate at the time of writing. Neither the author nor New York University Press is responsible for URLs that may have expired or changed since the manuscript was prepared.

Please contact the Library of Congress for Cataloging-in-Publication data.
ISBN: 9781479814398 (hardback)
ISBN: 9781479814411 (paperback)
ISBN: 9781479814435 (library ebook)
ISBN: 9781479814442 (consumer ebook)

New York University Press books are printed on acid-free paper, and their binding materials are chosen for strength and durability. We strive to use environmentally responsible suppliers and materials to the greatest extent possible in publishing our books.

Manufactured in the United States of America

10 9 8 7 6 5 4 3 2 1

Also available as an ebook

For C. Ray Borck

CONTENTS

Preface: The Legend of the Spider Goat	ix
Introduction: Meeting the Herd	1
1. Spider Encounters: Silking Comes First	31
2. The Gifts of Goats: Milking Them for All They Are Worth	61
3. The Goat As a System: From Pure to Artificial and Back Again	99
4. Thin Skinned: The Promise of Spider Silk Products	129
Conclusion: Knowing You're a Goat	153
Acknowledgments	169
Notes	173
References	191
Index	203
About the Author	207

PREFACE

The Legend of the Spider Goat

When I first heard that there were goats that had been bred to lactate the silk of spiders, I found myself obsessively googling news stories about them, trying to find out everything I could about how and why they existed. I read that these goats are able to lactate spider silk because they have been *transgenically modified*, which means DNA from spiders was introduced into goats so that their milk produced an extra protein. More than once while my daughter and I waited for her school bus during spring term did she nudge my arm because I was lost in my phone, piecing together facts about these animals. I was intrigued by what felt like a fictionalized science experiment of human modification of animals. I was captivated by the imagined futures of medicine and innovation these goats might promise—fabrics so tough and flexible they protect you from bullets, sutures so fine there is no scarring, artificial tendons and ligaments so strong our knees perform better than ever? The possibilities felt, and still feel, spectacular.

I wanted to meet these goats that produced milk containing modified silk protein. I was curious about how they came into existence. Did they look and behave differently from other goats? I conjured up freakish goatlike monsters tangled in sticky webs, all twitchy, bleating, and shuddering. Or I imagined superhero domesticated goat mash-ups spraying webs from their udders on command. I wanted to interview the scientists who created and raised the goats. How did this happen? Channeling some crunchy, animal-rights-loving earth mama, I thought I might give them a piece of my mind. And still I wondered how they were changing the future of medicine and the military. How do these goats fit into the larger moves toward transgenics and synthetics? What do they make possible?

In my milieu at the intersection of State and City public liberal arts academia, Brooklyn parenting, and middle age, I have found that my friends and colleagues usually expect me to join them in their disgust of scientifically engineered animals. Since I am a social scientist, a feminist, a mom, a queer person, I have felt encouraged to share feelings of horror at money-grubbing corporations exploiting all of vibrant life, squeezing the original purity from innocently portrayed domesticated animals. As recent attitude studies reveal, humans, especially in the more commonly studied Global North, are troubled by human intervention, particularly transgenic, in the natural world.[1] In a culture saturated with dystopian transgenic entertainment (think of film franchises like X-Men, Spider-Man, and Jurassic Park) and doomsday fetishization (the young adult Hunger Games series of books, the popular television show *The Walking Dead*, and Cormac McCarthy's book *The Road*), I can get swept up in the social pressure to be disturbed. Add a global health crisis, and it's off to the races of prognosticating apocalypse and doom-scrolling as dysfunctional lullaby.

Dystopia is not just a genre of media. Pandemics are no longer restricted to the realm of fantasy, as the reality is ever present and devastating. The contemporary cultural ambivalence around science, fueled by a sociopathic former leader, generates stories of "mad" scientists obsessed and giving in to obscene and unexamined psychotic urges to create monsters by either poisoning us with vaccines or creating chimeric clones of animal-human hybrids. I should pity (or educate) those duped by promises of a future where we are all fed and healthy, and we must remain fully aware that these scientists, real and imagined, breathe life into the dystopian future—as pressure from the moral chorus seems to indicate. And while some arguments are persuasive, I also resist the notion that humans' intervention into human and nonhuman reproduction is always nefarious. Humans do have multiple and flexible positions with respect to transgenics. In spite of the dominant, monolithic cultural perspective, I believe that my perspective is messier and less clear-cut and that power is expressed in competing and contradictory ways. Like semen banks that both undermine and support phallocentric gender roles, the semiotics of morality run horribly gray within this frame.[2]

Notes from a "Mad" Scientist: Identity and the Disintegration of the Natural

Technologies can be applied in ways that support liberation as well as domination. I have directly experienced the negative impact of laypeople's fear of technology. I have also used scientific innovations (however low tech, such as at-home insemination with a syringe) to reproduce. In my previous work, *Sperm Counts*, published in 2007, I "came out" about this practice. *Sperm Counts* is a sociological and qualitative analysis of multiple sites of physical and social contact with human sperm; I interpret data from sperm banks, sex worker interviews, children's books, and DNA forensic science and make connections between cultural beliefs about masculinity and socio-technical manipulations of sperm and semen. I am a queer mother of three daughters conceived through donor insemination using both known fresh semen and unknown banked technosemen modified by scientific manipulation, stored in liquid nitrogen, and shipped across the country.

The reaction to this book, while positive in academic settings, was varied in public forums. During the early 2000s, through radio interviews, public talks, and in the comments sections of popular media, I was called out for being a "man-hating feminazi," for "buying my kids at a sperm bank," for rallying for the end of men, and for being a "freak of nature." When I was teaching at the College of Staten Island more than fifteen years ago, some of my students asserted that my kids weren't really mine, because I was a lesbian who used donor sperm. A few even questioned whether my children could have souls.[3] Although things have shifted in the fourteen-plus years since *Sperm Counts* was published and queer reproduction has become more prevalent, contestations over the creation and existence of queer families persist. For example, nine out of the ten most censored young adult books are about transgender people, and the Trump administration worked to change the rules to end all LGBTQ asylum cases.[4]

Since my children and my domestic life are tightly woven into this ethnography of transgenic animals, it's important for me to introduce them, with their consent. My three daughters are Grace (twenty-three), Georgia (twenty-one), and Greta (twelve). They were conceived through intracervical insemination (ICI, sometimes called the turkey baster method) using donor semen, both fresh and frozen, from three men. These donors were not active parents in my daughters' lives.

My daughters were raised by several parental figures in addition to me. I gave birth to the older girls while in a lesbian relationship, so Grace and Georgia have two mothers. Their other mother has remained close to our family, living blocks apart since our break up several years ago. I gave birth to my youngest daughter while in a marriage with a transgender man; Greta has two parents, a mom and a dad. Additionally, she considers my former girlfriend a member of her family who is invited to every holiday and celebration. Grace and Georgia also see my former husband as their stepfather; he and I reside in separate apartments in a duplex in Brooklyn. For the past several years, I have been in a live-in relationship with a transgender man who is fifteen years younger than me. He is also a member of our family.

It is not my intent to commit some queer virtue signaling or to amplify my or my loved ones' "otherness" as a ploy. While I am not ashamed to share any of this, I feel it is risky but necessary. My work as an ethnographic and qualitative sociologist is situated and grounded in my very personal experiences of technologically enhanced reproduction and transgressions of normative gendered scripts. My embodied everyday life of being a white, cisgender, queer but heterosexual-passing, menopausal mom of three daughters (all technically half sisters) partnered with a younger transgender man intersects with the synthetic, the unnatural, and trans-ness themes that reverberate throughout this book and my analysis.

These embodied locations are my vantage points when considering a species like spider goats because of my personal life juggling the relatedness of my kids, the "naturalness" of our family, the incubation of biological experiments inside my body, and the cultural surveillance of my body, my children, and my family. Admittedly, I have been seduced by *genetic fetishism*, or the desire to have a genetic relatedness to my offspring.[5] I chose to make babies through mixing donor sperm and my eggs from my own body, rather than making kin in other ways. (The choice was partly driven by our meeting with an open-adoption attorney, who told my partner, a transgender man, and me that we would not be a likely choice, because of our queerness.)

My work on a meta level has always been about creation, including my kids. For some time, as a writer and thinker and mother, I have been creating objects that lift weight, and I am always putting these objects into contact. It is generative for me, and my embodied experiences as a queer parent are intrinsic to my knowledge production and my identity. My life is lived to subvert the notion of the natural and the constructed. In many ways, my life reconstructs the natural and at the same time foregrounds the constructed nature of everything we call natural. I am the "mad" sociologist, destroying nature itself, through actions and writing.[6] Sometimes I do experience "nature" as oppressive; it deserves to be reconstructed to better serve those whom it oppresses.

For example, as a feminist, a gender studies professor, and white woman in menopause, I see how the "worth" of animals, specifically humans (including myself), goats, and spiders, hinges on our time-delimited reproductive capacities. Over the past decade, I have conducted research about more-than-human relationships, as I am deeply persuaded by geographer Sarah Whatmore's claim that "humans are always in composition with nonhumanity, never outside a sticky web of connections or an ecology of matter."[7] Throughout this book, I draw on my lived experiences, my everyday life, to elucidate this stickiness, how I come to understand the transgenic goat. Like many before me, I cannot engage in a detached theorizing about nonhuman animals.[8] I prefer to conduct research and write in ways that echo philosopher Lori Gruen's entangled empathy, a critical attention to caring for all beings while recursively witnessing and responsively attending to the suffering of others.[9]

Real or Notional

In the late spring of 2017, just before I conducted the preliminary research for this book, I was in Dharamshala in Himachal Pradesh, India, for six weeks to teach sociocultural studies of food for Purchase College's study abroad program. The area is a mountainous region in the foothills of the Himalayas. Beyond our classroom work, we hiked (and scrambled over) several guide-led steep trails. The incredibly rugged terrain and high altitude caused even the most athletic among us to struggle at times. While each vista was astonishing, the required cheerfulness of motivating twenty-two mostly inexperienced undergraduate hikers (while navigating my own orthopedic challenges) was difficult, to say the least. Yet each time we encountered a herd of goats on a trail, it was a welcome relief. Their ordinariness in such singular circumstances made me feel like I could survive. Surely something so regular for them could also be regular for me. I'm a Capricorn after all, and like these goats, I could be sturdy, smelly, and dirty. Sadly for me, they were completely disinterested in being approached or handled. Initially I did not consider the greater sociological structure of these goats' lives, the flows of labor, capital, colonialism, migration that co-constituted goats with humans mediated by ecologies, legacies, histories, and climates. I thought of them as some unadulterated representation of the bucolic and an example of some purer form of life—more goaty than American farm goats. As a white American teacher-tourist, leading other Americans on an international experience to a place I had never been, I considered the goats just another object to authenticate this experience; they were parts of digital photographs and embellished stories to be exported and later shared about the "real" experience of this part of India. Only later did I consider the relationship goats have to security and livelihood in the Himalayas—ideas that went beyond my American postcolonial romanticization of the goat and, by extension, the people and the landscape.

The region these goats live in is one of great mobility for humans, other animals, and weather fronts.[10] For example, Dharamshala is the site of massive migration for Tibetan refugees who have (for decades) followed the exiled Dalai Lama to India. Though 20 percent of the world's goat population is in India, goats are not native to this country; they migrated with humans from the Fertile Crescent more than ten thousand years ago.[11] Goats have been integral to the local economy for centuries, but the sustainability of goat herding as viable economically, logistically, and materially is under threat because of climate change.[12] The tree line and snow line are rapidly changing with a 1.6 degree Celsius increase in the temperature in the past decade.[13] Additionally the goats' health is endangered by an increase in parasite infections from the changing environment. The promise of stable goat-human collaboration, relied on for generations by both species, is now uncertain.

Figure P.1. Goats on path in Himachal Pradesh, India, June 2017. Used with permission from Anaïs Baptiste.

Clearly my perspective on the goats from India, while always partial, became more evolved and complex. But what is remarkable to me is how my initial understanding of the goats relied on some naive notions scaffolded on nostalgic colonial fabrication about the essence of goat. My own American, white, and female sense of self structured the way I perceived *goat*. If those Himalayan goats performed the work (cultural and material) of real goats, maybe I viewed other goats, spider goats, and farm goats as some type of notional goats, as simulacra.

A few weeks after returning to the United States exhausted, I flew to Utah to meet a very different herd of goats. These goats were more uniform in appearance and housed in separate pens in an administratively orderly and purposeful fashion. The herd greeted me by walking up to the edges of the fencing and nibbling on my jacket or my fingertips. The

Figure p.2. Goats at South Farm, Logan, Utah, July 2017. Photo by Lisa Jean Moore.

animals were eager for touch, good-natured, seemingly curious about me, and almost friendly. And although I remembered the Himalayan goats from just weeks before, I wasn't soothed by these goats despite their amiability. My knowledge about their circumstances made our meeting slightly eerie.

These were not ordinary, or even notional, American farm goats. Rather, this herd in Utah has been bred to contain the DNA of spiders and to produce spider silk protein in their milk. Spider silk protein forms the material that spiders use to build their webs and other structures. The mechanical properties of spider silk are essential for the life of the spider, and spider silk is a highly promising material for use in the fields of defense and biomedicine. Its uses include parachute cords, tire linings, high-performance sportswear, and medical applications such as replacement ligaments, tissue scaffolds, drug storage matrices, and drug delivery systems. Although spider silk holds great promise, it is difficult and impractical to harvest in its natural form because spiders are territorial and cannibalistic. The scientific product from generating and harvesting the silk protein of this spider goat species, made possible in part through funding from several branches of the US military, has been called Bio Steel. The ubiquitous claim of spider silk is that it is stronger than Kevlar and that this new lightweight, ultrastrong fabric could surpass Kevlar in protecting soldiers and police from shrapnel and bullets. Researchers also hope the spider silk will revolutionize medicine with biocompatible medical insertables such as medical adhesives, prosthetics, sutures, and bandages.

Scientists can artificially incorporate genes from one organism into another, a technique most commonly associated with the genetic modification of such crops as corn, rice, soybean, and cotton. This process of manipulating the genetic material of an organism to include the DNA of another, unrelated organism creates new transgenic organisms. To date, several mammalian species—including these goats as well as mice, sheep, and cattle—have been genetically modified to be used as research objects or to produce biomedical devices and other biomaterials.[14] According to the US Food and Drug Administration (FDA), transgenic science is also used to make pharmaceuticals, particularly if the animals are biocompatible with humans. Transgenic science is rapidly becoming a source of proteins, cells, tissues, organs, and materials common to the practice of medicine. Therefore, these goats are semiotically and materially different from those in the Himalayas. Each herd is layered with meaning, co-constructed in a relation to humans, and valued for the use they can be to humans. But there are important qualitative differences that I aim to explore in this book.

Introduction

Meeting the Herd

I had finished my previous book on horseshoe crabs when a friend suggested I look into spider goats after she had read about them online. For my friend, the fantastical idea of altered lactation is what drew her in, but for me, the hook was far less dramatic. I have long had a fondness for goats, and my work with horseshoe crabs fostered a sense of responsibility to invertebrates. Spider goats brought these two together. As I read about them in my initial research, I found that humans had genetically modified some goats to create systems that synthesize and express spider silk protein. This protein can then be reconstituted to engineer military-grade fabrics, biomedical devices, sporting apparel, and luxury face cream. The spider goats are a tool for producing this protein on a larger scale—something spiders cannot reliably do.

A binding theme of this book is the examination of interspecies relationships between humans and goats and spiders—our multiple sites of entanglement and enmeshment are always becoming what and who we are. My research exposes how humans, goats, and spiders connect with each other in highly interdependent webs. Additionally, this is a story about how spider silk—the product and the hopeful promise—has yet to be delivered to the market. As the expense of maintaining transgenic animals ratchets up, the clock is running out on their viability as a system to effectively produce the protein, a protein with no consistently profitable commodity to justify its considerable investment. I explore the possible obsolescence of the goats (and their possible extermination) alongside notions of disposability in aging technologies (myself included). *Our Transgenic Future* examines how our human vulnerability offers opportunities for scientific innovation and profiteering from transgenic species, revealing the stakes of transforming animals into capital. I suggest that animals here (in this spider silk context) are capital in the classical and neoclassical sense. They are commodities (the spider goats themselves), laborers (that make spider silk and kids), and raw materials or assets (in the sense that they produce more transgenic goats and have value for their modified capacities). But spider goats also join other farm animals as emergent forms of capital in a dynamic biotechnological, transgenic field.

Anthropologist Sarah Franklin's groundbreaking work on animal cloning explores how centuries of sheep breeding combined with innovations in cloning technologies birthed a multifunctional resource—cloned sheep.[1] Like sheep, goats are both old capital livestock and new capital as vehicles for biocommerce. Alex Blanchette's sociological ethnography of agribusiness's modification of the hog demonstrates how the extraction of value from the animal has fueled the inevitable human reliance on hog commodities in everyday life (through food coloring, adhesives, biomedical products, and biodiesel).[2] Additionally, these modified animals, systems for expressing synthetic products, are a form of lively capital, a convergence of life sciences systems with regimes of capital.[3] Ethnographic methodology enabled me to observe and analyze everyday microinteractions between scientists and students, humans and goats, spiders and entomologists. In the field, I saw, and now describe here, how these encounters were produced, structured, and normalized by means of capital's production of spider goats for the physical extraction of their profit potentiality. These living animal machines worked side by side, cooperative and obedient, the herd of them, a laboring factory.

Since this ethnography traverses several field sites and literatures, I hope that this book will reach a transdisciplinary audience. For me, it has always been a bit difficult to know how my work will travel and who it will move. *Catch and Release*, my book about horseshoe crabs, in part inspired a documentary filmmaker to create *The Whelming Sea*.[4] The topic of this book is difficult to put into one field of study. I am a sociologist but not very disciplinarily pure. I am a feminist, but my commitments are sometimes questioned. Some interdisciplinary scholars may find the book useful, since it involves critical animal studies, science and technology studies, gender studies. I am also speaking to ethnographers and, more broadly, students of qualitative methodology, as I am using my literal self as an analytic tool to understand the goats, the spiders, and the scientists. And while I want to contribute to a broader project of critical inquiry, this book evolved from my desires to learn, create, experience, and share, rather than my attempts to intervene or advance.

The beauty of ethnography, when done well, is its documenting of a lifeworld. I am attempting to document the lifeworld of humans, goats, and spiders organized around a particular technology over one period. In his multispecies work on wild horses in Spain, anthropologist John Hartigan writes, "Ethnography is a means of attending to and analyzing the plethora of instances and interactions that escape even a rigorous quantitative approach."[5] I try to describe this interaction of mammal and invertebrate species creatively, but as a sociologist and an academic, I am not outside disciplinary socialization. I have been well trained. Like the spider goats that are milked twice daily and that wander a little bit before they walk back into their pens, I do some gentle wandering before I come back to my pen. I don't really know how to totally leave sociology or animal studies or gender studies—I have been successfully disciplined by these, but I do a little meandering too. It is productive and it changes what I think, what I do, and who I am.

Multiple Eyes: Interrogating Meanings at the Periphery

Feminist work that embraces the interrogation of the self, the body, embodiment, and fragility as units of analysis has always captivated my imagination. It is this type of work I hope to generate. I take as one model the work of interdisciplinary artist Kathy High. High interweaves a story of mounting an artistic installation of laboratory rats with exploring her own autoimmune disease and interspecies connection to the rats.[6] These three rats were all transgenic, microinjected with human DNA to exhibit rheumatoid disease. High purchased the retired rats and designed a physical enclosure and a guided museum tour as an art installation to examine the role of transgenics in human and nonhuman animal life. The written piece is deeply moving at conveying the affective stakes of connecting with nonhuman animals while trying to provide a more gentle everyday life for exploited animals. How do we, as scholars, scientists, and laypeople, pay attention to the very intimate entanglements and materials and meanings we produce with real live animals? As I approach my own fieldwork with animals, I can sometimes soothe myself with my very hands-on petting of goats, horseshoe crabs, or spiders. I am being gentle with them, after all. But I can almost forget that my own livelihood and health are only made possible because of their milking, bleeding, dying. I am already and always entwined with the goat even as I first pet her head.

Painter Sunaura Taylor fully integrates her subjectivity into her artistic production and critical writing.[7] Her subjectivity of creator (painter, author, researcher) and her relationship with other beings generate her feminist technoscience practices. Documenting her consciousness transformations of her childhood, Taylor, a disabled person, was attuned to the suffering of animals from a young age and intermingled these feelings with her subsequent self-identification as a "crip." Her lifelong practice of painting representations of animals is informed by her experiences of ableism, which she defines as "how we define which embodiments are normal, which are valuable, and which are 'inherently negative.'"[8] Ableism is the alibi used to coerce, subjugate, and enslave people with disabilities. What is so important about Taylor's writing is her ability to put herself into the implicated positions as she reveals the details of her relationships with Bailey, her small mixed-breed service dog, which becomes physically disabled. Negotiating the interspecies relationship and how Taylor becomes a service human to her dog brings to life the epistemological terrain she offers in the book. I strive for this type of subjective entanglement during my research.

Interweaving stories of the self with interpretations of the ethnographic data adds productive layers for comprehension. This precise blending of personal stakes and autobiographical revelation brings research projects to life. Sociologist Alondra Nelson combines her own personal history with the history of DNA in African American root seekers and reparations.[9] Starting with her family's viewing of the television miniseries *Roots*, Nelson traces her journey of genetic genealogic discovery alongside her examination of the rise and use of DNA testing services for reclamation projects.

Beyond autoethnographic-adjacent scholarship, I am inspired by the work of feminist technoscience scholars, in particular, the work of the sociologist Ruha Benjamin, who uses fiction derived from methodologically sound empiricism to push our epistemological capacities. A few years after Benjamin's meticulously researched *People's Science: Bodies and Rights on the Stem Cell Frontier*, she wrote an article that, through fiction, speculates about the possibility of pushing her sociological imagination to reconfigure race, science and subjectivity.[10] Benjamin's work offers "new modes of representation and engagement that exceed the traditional bounds of academia—expanding what counts as knowledge." While my work is not fictional, I do integrate an assiduous and mediative reflexivity as a way of pushing my own thinking and visualizing the linkages between ideas and themes.

As feminist researchers, we choose to study ideas based on our oft-acknowledged positionality and emotional investments in the world. In studying the tenuous journey of a pharmaceutical start-up, Anne Pollock, an anthropologist of science, acknowledges being compelled by the mission of the company (to develop drugs by and for South Africa) and shares her realization that she was "studying up."[11] Beyond what we study, positioning ourselves as reflexive feminist technoscience scholars also means we are *asking different types* of questions, as suggested by reproductive neuroendocrinologist Deboleena Roy.[12] Building from Sandra Harding's standpoint theory, an intellectual perspective that claims knowledge comes from a social position, and laying bare the limitations of the standards for achieving objectivity, Roy calls for a use of standpoint theory to prescribe new practices for generating knowledge. The standpoint of a feminist scientist is that of a marginalized knower, one who, according to Roy, can critically see dominant ideologies embedded in scientific theories, practices, paradigms and languages. As an outsider within, Roy describes her practice of science as being inextricably mediated by her standpoint as a woman and a feminist; she analyzes her decisions not to work with live animals and her favoring an in vitro approach to discovery as part of her doctoral research project.

While persuaded by this argument, I do wonder what is the standpoint when a human is trying to understand the experience of the animal within a fieldwork site. I suppose that because of my standpoint as a lactating mammal, I am asking different questions about transgenic goats that lactate spider silk protein than, say, would be asked by a scholar who does not have this standpoint. My standpoint as a mammal who has lactated is an asset to my knowledge production (one that I mine in chapter 2), while at the same time, I fear the essentialism it flirts with. Does this mean that some nascent solidarity I have with the goats enables different questions to emerge?

Sharing my own reproductive stories (either my own or now those of my daughters) has often meant getting interrogated by others about our realness or naturalness as a family. "Who is your real mom?" "What do you mean you don't have a dad? That's impossible." My body and my reproductive intentions are suspect; I was the incubator of unnatural children (despite my own genetic fetishism). These quizzes about my reproductive behavior have underscored how questions about naturalness often imply a certain morality and normative judgment about social arrangements. I hope that I am attuned to the ways that claims of the natural can be used to force people, animals, and social relationships into alignment with normative and structural dictates.

There is wisdom in mining the stories of human reproduction, including my own stories, to remember how "fantastic creatures" become ordinary or normal over time. Advances in assisted reproductive technologies have meant that in my lifetime, the so-called test-tube baby went from freakishly queer to incredibly common. Now for an increasing number of women who rank high in systems of stratification, these procedures are unquestioned. Where a test-tube baby was a fantastic creature, my children seem rather ordinary. Even when I discuss this project with others, although a vast majority of people say, "*Spider goats! Are you serious?*," a steady stream of others shrug their shoulders as if transgenic species are rather passé—they've heard that one before. The paternalistic thrust of institutional scientific prerogatives transforms goats' duties beyond being a food or clothing source to produce an additional protein in the lab—spider silk. But what level of goat modification becomes unnatural? And for how long is it considered unnatural? What are the Baradian cuts we are making?

My work is influenced by the scholarship of Karen Barad, particularly her insistence that we attend to where we make decisions in our discursive renderings of nature.[13] Barad says that when we are observing something in the natural world, we do so to acknowledge where we make the cuts. Differences materialize because of how we make these cuts. We are continually making decisions when we do science, observe natural phenomena, or study animals. There is a difference between a Himalayan goat and a spider goat because I make cuts around geographical location, genetic manipulation, and sensorial interactions. I then create dichotomous categories to represent and produce material reality. The reality is both material, insomuch as something is physically happening (goats are scrambling over mountains or wandering in pens), and discursive, in that we come up with words, definitions, and descriptions of how to represent the materiality. Discursively, I believe that the Himalayan goat is meaningful because it is pure and natural, and I exclude observations and interpretations of these goats as dipped in chemicals for parasite removal, fed antibiotics, or reproductively managed for humanistic breeding outcomes. I privilege some notions of their indigeneity, relying on some version of my own racially skewed notions of wildness and native origin. I work to attend to this traffic between my own geographic, biographical, and structural positionality and my beliefs about goats.

Fragility of the Natural

Human questions about what is natural and what is altered have always been in a tense relationship; they are a contradiction without a coherent narrative. Our biological understandings of procreation, the facts of life, are the cosmic origin stories of Western culture. Our scientific conceptualization of coming into being is assumed to simply be "nature stripped of all its cosmological moorings and therefore presumably generalizable to all people."[14] With the innovation of new technologies and the reconfiguration of sexual identities, gender expression, and family compositions, there are alterations in how we (as people, animals, plants) come into being. These emergent forms challenge the basic foundations of our understanding of ourselves and the world around us.

My conceptualization of female goats as malleable surrogates to produce a transgenic material that will better the human race comes from feminist scholarship about human women's reproductive capacities linking them to animality.[15] I see connections between the use value of a female human body and the female goat in that they are reproducing and nourishing generations. The proliferation of the processes of surveilling, monitoring, and innovating biological reproduction is gendered and racialized. The interventions to the female human body are racialized, whereby female bodies are (re)producers and where "improvements" are designed so she (human) can produce (fulfill biological destiny) for the human race—producing, depending on her racial identity, either children or labor. If we liken women to other animals, they can then be objectified and modified to be better reproducers, or better breeders, depending on the relative worth of those women. Historically it has been mostly white, male scientists (mostly paternalistic institutions) "helping" white women accomplish their duties, while other women, especially Black women and other women of color, are simultaneously rendered as less than human, closer to (or less than) animal and more aggressively exploited.

In other words, female reproductive capacities, particularly in the US context, come from a history of racial capitalism, reproductive slavery, and the continuance of both processes in contemporary biocapitalism. As Alys Eve Weinbaum's *Afterlife of Reproductive Slavery* examines, the four centuries of breeding enslaved people have shaped human reproduction and created circuits of exchange that enable the commodification of women's labor power—eggs, babies, and reproductive labor.[16] In the case of human reproduction, Black women's bodies continue to be the site for massive biological and political intervention as Black women were constructed as nonthinking subjects akin to animals. In the United States, white women were invited to become thinkers, and therefore human, to the exclusion of other women.[17] Clearly, humanization is racialized; not all stories of becoming human are homogenous.[18]

A product of my liberal education, I had naively thought several years ago that liberation projects that enabled white women to become human must be pursued for all groups of people. However, it is not enough to just offer an invitation to humanness, extended to Black women as a reparation of this problem. Scholar and cultural critic Zakiyyah Iman Jackson acknowledges the Western tradition of linking Blackness and maternity with animality. But more significantly, her work, a literary criticism of diasporic Black authors and artists, brilliantly shows that to be placed in the category of human does not necessarily offer protection against anti-Blackness. From her perspective, including Blackness in a universal humanity is to refigure Blackness as always abject or Blackness as "the animal within the human." *Human*, as a category of belonging, is a violent imposition of a racially stratified conception of humanity.

Jackson's work has forced me to reconsider new hierarchical rankings. Significantly, in many ways, the carefully guarded, meticulously tended lives of both the natural and the bioengineered goats diverge from Black and other marginalized people's motherhood in that the goats have more resources dedicated to them than do many of these mothers. Jackson's scholarship forcefully works to disenchant people (including me) from the idea that humanization will lead to greater equity.[19] As she recently stated during an interview, Jackson wants readers to move away "from a melancholic attachment to what an ideal human is."[20] My "melancholic attachment" to pure goats and human neoliberal projects is disrupted.

Just as many might be suspicious of transgenic innovation and wish for a protection of the "pure" goat, there is also an active cheering squad for this type of scientific progress. In recent academic and scientific literature on transgenic species, some humans are remarkably celebratory about innovating these new and improved nonhuman animals. There are, for example, the dairy cow embryos changed by human genes to lactate human breast milk; a transgenic bull modified with human DNA to reproduce modified dairy cows; AquaAdvantage salmon, modified with genes from an ocean pout for faster growth; and Enviropig, transformed by the introduction of a transgene created from the bacterium *Escherichia coli* (*E. coli*) to produce less waste.[21]

But these promises of transgenic creations saving us from waste or starvation can create new problems. As sociologist Jonathan Clark, who has written extensively on the Enviropig, points out, transgenic animals can often become surplus when they are no longer deemed useful to humans.[22] Too expensive to maintain because of biosecurity anxieties and with limited value, transgenic animals, as we will see with spider goats, have a hard time justifying their existence (sometimes uncannily familiar to my experiences convincing people I wasn't doing harm raising my daughters without a genetic father). The value of transgenic animals is usually defined by their value to human commerce. As they become less useful, some animals are allowed to be retired, whereas others are killed. I will return to this notion of what makes spider goats killable or, perhaps more passively, what allows them to be left to die, in chapter 2.

To many, genetic modification is a form of contamination to the presumed purity of the natural. Studies have demonstrated that people think genetic modification is a challenge to their moral values.[23] Echoing the foundational work of anthropologist Mary Douglas, laypeople view naturalness and the natural as sacred, unnerved by the specter of genetic modification and its capacity to pollute—genetic modification arouses our moral intuition as purity is violated.[24] It is clear that naturalness is a subjective and partial state, whereas a person who opposes genetic modification can still rate domesticated animals (also known as *human-caused genetic selection*) such as goats or dogs as natural while expressing high disgust at even scientifically proven "safe" genetic modification.[25] Significantly, the transfer of DNA between different species (as in the case of spider silk protein introduced to goats) is less acceptable to people than is transferring genetic material within a species.[26]

These types of public opinion studies typically lump all transgenic organisms together rather than assigning them different categories. In a large representative study of six countries (France, Germany, Italy, Switzerland, the United Kingdom, and the United States), people were asked their opinions of genetic modification.[27] Their responses were negative in all countries and more negative in continental Europe. Millennials' attitudes about genetically modified foods are similar to those of the rest of the population, although millennials with higher education are more willing to purchase genetically modified organisms than are their peers with less education.[28] Opposition to genetic modification is not affiliated with any political ideology.[29] Concerns about human health, human safety, and animal suffering are often cited as contributing to negative attitudes about genetic modification.[30] Despite ambivalence about genetic modification, politicians around the world often advocate for genetically modified organisms as viable solutions, as exemplified by a British Parliament member's statement in 2012: "The resilience we need for the future will be delivered by smart plant breeding—and that's all GM [genetic modification] is."[31] But this celebration is tempered by cultural anxieties about manipulating the basic reproductive mechanism of other organisms and the longitudinal effects of changing "pure," natural forms in these ways. Despite consistent opposition, humans are still transgenically modifying animals. How might we explain our own species' contradictory behavior?

Weighing Risk

In terms of moral intuition about purity and contamination, many people may believe that when nature is enhanced or modified, a level of unacceptable risk is introduced. Even with longitudinal data, we cannot know all the risks lurking in the future for the transgenic animal. And although humans negotiate risk every day, the idea of engineering risk into an animal, for some, may be a bridge too far. But what do we make of the industry tagline that producing biotechnology is a way to improve life and produce a more predictable, safe, and bountiful life? Genetic modification is often introduced and touted as a way to ameliorate or eradicate an existing or anticipated risk (risks that are very real or vividly imagined). We must weigh risks whenever we introduce technology into our everyday lives, for example, when we get a vaccine or give a child a cell phone. For example, when I was picking technosemen for my own insemination, I experienced this dizzying anxiety of navigating present and future risk participating in biotechnological choices. Which donor profile was the best? Which types of semen vial would work most successfully—washed or unwashed, fresh or frozen? What about the social, emotional, physical, and genetic health of this future child?

The scholarship on risk by European sociologists Anthony Giddens and Urlich Beck offers an angle to interpret the case of genetic modification in general and the creation of spider goats in particular.[32] We live in a risk society, which Giddens describes as "a society increasingly preoccupied with the future (and also with safety), which generates the notion of risk."[33]

According to Beck and Giddens, our modern lives have become structured by our ongoing risk assessment at both the individual and the institutional level. Instead of accepting that risk is just a part of everyday life, we now see risk as part of human action. Therefore, the threat of risk (or bad outcomes) is now something we can control and anticipate. As modern humans have moved away from predestined futures, we must hone our ability to assess risk or amplify our awareness of risk to live as modern subjects. Giddens likens risk to an "energizing principle" that is linked to productivity in industrialized modernity—we can make new things to mitigate or manage our future risks.[34] It is good business for us to be attuned to risks as potential consumers of solutions.

Interdisciplinary scholars have investigated how discursive and material forces of modern-day capitalism manufacture risk. Merryn Ekberg, a health sociologist, explains how a risk society emerges through the "collapse of inherited norms, values, customs and traditions," whereas "the risk society is characterized by dislocation, disintegration and disorientation associated with the vicissitudes of detraditionalization."[35] So while the move away from strict adherence to heteronormative gender roles or prescribed religious rituals can be liberating for modern subjects, the lack of clear social rules may also cause anxiety, disenchantment, alienation, or fear. And humans often communicate their risk assessments to each other through social media.[36] Living in a risk society means we also advertise risk to one another; we are increasingly alerted to, and pass on alerts about, existing and new threats such as viruses, food recalls, computer hacking, identity theft, sex trafficking, carjacking, sea level rise, and domestic terrorism.

The proliferation of human-made risk slowly erodes public trust in social institutions, including science and medicine, as we see so clearly with the increasingly devastating effects of climate change or the public's mistrust (fueled by political theater) in lifesaving technologies like vaccines. This modernist turn pushes us to develop a keenly honed inward reliance—a self-determinations of what is risky and what isn't. No longer trusting experts, we scramble to take in as much information as possible and then are forged as our own experts. Without the broad-ranging support of social institutions, individuals become fearful of the innovations that come from scientific innovation like synthetic biology. Ironically, synthetic biology offers new products to manage risk but faces an uphill battle of convincing fearful potential consumers. Risk is discussed further in chapter 4.

Methods

Social scientists examine these shifts in scientific innovation and cultural integration of new realities like the cloning of Dolly the sheep, the proliferation of genetically modified organisms, and the expansion of assisted reproductive sciences. This manipulation of the natural environment is occurring at a time of massive ecological degradation and destruction. The boundaries of life are more porous and hybridized. Or perhaps we just *feel* as if life has become more porous and hybrid. I think it's a more robust sociological approach to think of positivist claims about what is unique about the present as embedded in discourses, ideologies, and anxieties, rather than as singularly universal truth. In other words, as the manipulation of the environment increases, sociologists study these changes. I do this by analyzing intersubjective discourses, not objective truths.

In this book, I structure this technical and scientific exploration (and its cultural reception) alongside and interwoven with my own perspective, which is informed by my reproductive practices. In my other books, I have also used experiences from my intimate life together with my professional and intellectual training. This form and the content of my reflexivity reveals my commitment to demonstrating how who I am is wholly integrated in what I know and how I interpret information. My work traces the evolution of silk milk transgenic production in terms of human concerns and personal reflection—sparingly but deliberately used. I share my personal experiences to demonstrate that thinking about the mundane is entwined with the extraordinary and emergent. There is no clear line between the dense science of engineering new species and creating novel products and the very human scientists who engage in this work or who study and interpret that work, as I do.

While my life may be part of the mundane, I don't distinguish it from the bigger scientific practices or forces. On the farms and in the labs, I am rediscovering mundanity as well as moments of wonder and speculation and boredom.[37] My personal reflections and lived experiences, and those of the scientists I am studying, are part of the same soup. I ground my interpretation and evaluation throughout this book in my humanness, however mediated, and in my femaleness, particularly in the degree of emotional work I continue to perform as a fieldworker and an empathetic knower. This project brings together all the things that I have worked on during my life. The entwined helixes of lived experience and academic training are the building blocks of my epistemology.

My analysis is based on a multimethod study that focuses on interviews with more than twenty-five people, including biochemists, engineers, conservationists, field biologists, veterinarians, regulators, and entrepreneurs, combined with three-plus years of fieldwork. I observed participants at the Spider Silk Laboratory, an academic center for spider silk experimentation and product innovation, and at South Farm, where the only living herd of spider goats was located, at Utah State University (USU). I also joined an entomologist during a collection expedition in Gainesville, Florida, and I interviewed several arachnologists.

To start this project, I wrote a letter to Randolph "Randy" Lewis, the principal investigator at the Spider Silk Lab, and sent him some of my previous work on horseshoe crabs and bees as part of explaining to him my reasons for studying spider goats and scientists like himself. I requested a phone interview, and after this initial contact, he was enthusiastic and invited me to come to Utah to see the goats. With this connection, my research began in the summer of 2017 on a preliminary fieldwork trip to USU to visit the Utah Science Technology and Research Initiative (USTAR) and Randy Lewis's Spider Silk Lab. USTAR provided the start-up funds to outfit laboratories for the spider silk project. The laboratory, located in Logan, Utah, is part of a research park established in 2006 by USTAR and originally served as a biotechnological incubator for the state of Utah. During the following three years, I visited USTAR one additional time and spent time with Randy and the other scientists and assistants working with him.

My research draws from qualitative methods and grounded-theory methodology.[38] I use traditional ethnographic data-collecting methods, including participant observation; intensive interviewing; observations of, primarily, goats and scientists in Logan and New York City and spiders and entomologists in Gainesville. Qualitative research is emergent and sometimes messy, as it focuses on meanings, interpretation, rituals, and interactions. In this context, I did not start with a hypothesis per se but prepared a set of ideas and questions that guided me through the initial phases of the project. As is common in qualitative research projects, these questions were modified and amended over time to reflect my interpretations of the ongoing fieldwork process.

My methods are inextricably linked to my development of nascent theories. Admittedly, my methods might not neatly fit into the discipline of sociology. I want to be very explicit: my observations are an epistemological and methodological move toward an even more micro, almost quantum, level where my interpretation includes feelings, discussions, relationships, affects, and sensations. My use of reflexivity and tenderness to its messiness is not new.³⁹ Feminist qualitative researchers grapple with how to navigate the embodied self in a variety of field sites.⁴⁰

But reflexivity is more than personal. All theories are surely intersubjective and built over time through a community of people. We cannot escape that subjectivity. This book proposes to rip apart the taken-for-granted notions of identity and how it affects scholarship and research. In all my work, I question how I situate myself. This questioning becomes a process in my fieldwork; my epistemology evolves over time and is based on enactment, not solely upon static characteristics. As I discuss in this book, for example, the memory of soreness in my own breasts enables me to have empathy across a species divide—my own identity and embodiment and my interrogation of it fuels my insights, perspectives, and theoretical contributions.

Logan, Utah

The fieldwork, primarily conducted in the laboratory and at the agricultural facility where the goats live, also took place in Logan, Utah. I didn't set out to do an ethnography of a college town in northern Utah, but I observed many things in the study of the goats, and it was striking that this location would be the context of transgenics. As a lifelong New Yorker, I was seeing "natural life" at work, and it felt and looked delightful.

At first, Logan is hard for me to apprehend, and I feel as if everything is cinematic; I expect to see tumbleweeds roll across the vast highways. The landscape is so flat it's as if you are looking through a wide-angle lens. I rented a big Chevrolet Silverado pickup truck (it seemed like the right choice). Each day, I drove between my Airbnb (an overly spacious guesthouse on a sprawling wheat farm with a little creek where a muskrat family swam in the early evenings, seeming bucolic and leisurely), the lab, and the goat stables. Driving between any of these points took about twenty minutes, not including trips I would take to the warehouse-sized grocery store or the state-sponsored liquor store. Thus, I experience the landscape primarily through the lens of a boosted big-screen windshield.

In Brooklyn, my commute to the college is a gauntlet of urban obstacles on narrow, crowded streets and highways filled with pedestrians, traffic, elevated train tracks, speed cameras, construction chaos, and potholes. Turning onto Atlantic Avenue every morning, I take a deep breath, crack my neck, and tense my shoulders, summoning the adrenaline necessary to drive the thirty-seven miles through Brooklyn, Queens, and the Bronx to the campus in Westchester in fifty-five minutes. Conversely, driving in Logan at the basin of the Cache Valley proceeds at a kind of even, lawful pace, with no tailgating, no weaving in and out of traffic, no red-light running, no aggression. The entire place appears as if in a dream state, and the ease of movement feels as if I could do it all in my sleep. Horses lazily turn their heads to watch me drive by. The streets are laid out in a grid of numbers and cardinal directions. The corner of E 1800 N at N 600 E.

Unlike Kings County, Brooklyn, there is no premium on space in Cache County. The roads are wide, the parking lots feel like small towns, the stores are warehouse-sized, the sky expansive. Beyond that, the farmland extends for miles and miles before running into the base of the mountains. It feels as if the landscape is hung inside a giant invisible hammock held up by those mountains, as if you're in the shallow section of the earth.

In Logan, everything feels easy and gentle, like the small waves of breeze along the surface of the golden wheat fields. Being in this environment has a deep effect on me and my sense of self. As a younger, queerer sociologist, academic, and urbanite, I was almost required to be suspicious of wholesomeness. I grew to resent the way the idea of virtuousness and health were used as moral barometers of social or personal worth. But as I witnessed the proprietors of my Airbnb on their daily early-morning, multigenerational march from the raspberry patch with three kids, two parents, and two grandparents carrying multiple pints of berries in flats, I eavesdropped on their conversations about planting potatoes and picking berries. Despite my tendency to lapse into the-grass-is-always-greener, I really believed they spoke with such sincerity, no harsh tones of malice or sarcasm.

As I reflect on this later, writing in my home, it is clear to me that I can get carried away by the cinematic feel of my surroundings. Leaving cultural stereotypes aside, it is hard to imagine that humans live lives completely devoid of tensions, contradictions, and conflicts as one might suspect; my mind carries the cultural juxtaposition and contrast of urban and rural communities too far.[41] Am I guilty of falling into the trap of Western anthropologists in the 1950s who were so taken with the differences of the societies they were observing compared with their own backgrounds that they missed the dynamics of the communities they were observing? In my idolization, I may have made it hard to see the weirdness of humanity there. I make cuts about Utah, Logan, the people but there is something very queer about all this normativity. And even as I admit to my own construction of false dichotomies, I recognize that some qualitative differences make me think differently in Logan—the goats are part of this landscape, the wholesome swaying of time and space, and yet also a queerly created synthetic machine.

My own internal conflicts can interfere with my ability to see clearly, as I feel so envious and regretful that I never had this type of cohesion and conversation in my own family with my kids. My daughters grew up in Brooklyn, in two different households, with many parental figures. Being here, I feel more gentle in my assessment and see that I am envious of this host family in some ways. And if I am honest, I'm a little jealous of the wholesome, heterosexual reproductive cohesion in their "intact" family as they talk about the land and walk back to their modest house. I am up on the deck, looking down on the scene as I write a review of a theoretically dense, indeed esoteric, text on gender, sex, and sexuality in animal studies. Some of the pieces are so obscure and insane. I default to my ordinary mantra: What am I doing? How is this my life? I miss my kids. They are dispersed in North America right now, not really integrated with me at that moment. What would the Freaky Friday situation be like if this family came to an Airbnb that we hosted in Brooklyn?

And yes I know, I know, my queer friends, colleagues, and even possibly one of my kids would tell me that this feeling is just my own false consciousness or my being brainwashed by the hegemonic representations that my hosts embody in their walk at sunrise from the raspberry patch—literally amber waves of grain behind them with purple mountain majesties as the backdrop. But maybe there is something to mourn in never having had or given this? There is some grief in my heart. And besides beating myself up about it—which I do often—I worry that my children will resent me because of it. I worry that my reproductive choices have both made them and at the same time taken something. from them.

My prevailing research method is to tease apart the cultural black and white when it comes to understanding themes. Utah versus New York; "natural" versus "synthetic"; sperm ejaculated from a penis inserted into a vagina versus sperm ejaculated into a cup, shipped in a tank, and inserted with a syringe; nuclear versus alternative family; red versus blue state; Mormon versus everyone else. I'm not a self-righteous dyke with no regrets, shame, or concern about using sperm donors. I don't think transgenic goats are the worst things in the world. I'm (not) sorry my daughters don't have traditional scripts—are they transgenic? And part of the lesson of this ambivalence is that the easy theoretical lens, like the cinematic lens I so often saw Logan through, for this project would be one about an absolute and assured ethics. But that reading is too easy, and it's not new, and it's not true to me or my methods.

USTAR Labs

The Spider Silk Lab is housed in a building constructed with USTAR money and embossed with USTAR logos and branding. As I pull up to the lab, I'm nervous. Having read about Randy and exchanged emails with him, I have a sense of him as direct, quick, and responsive, but I still have doubts that he'll grant me access to the lab. I climb the stairs and wander around the office space, knocking on his door. There he is at a standing desk, seemingly ready to multitask. I do a double take as I notice two of my books on his large desk. He immediately suggests that I check out the lab and walk around, assuring me that everyone is aware of my visit and that I can interview anyone. Just like that, I have been given full access and welcomed warmly.

I feel comfortable as I explore the laboratory for the first time. The physical layout is spacious and bright with ample natural light. Quirky, humorous signs are pasted on the entrances to the walk-in freezer and refrigerator, Spider-Man kitsch—blow-up toys, stickers, action figures—decorate lab spaces, and hardware trays of socket wrench sets, drills, and pliers fit into corner spaces. The lab has a fun, playful, mischievous feel to it, with various tinkering projects laid out on work benches.

In the Spider Silk Lab, there are four systems that create spider silk protein.[42] A system is a bioengineered organism that has been modified to make the protein. The four systems in this lab are the goats, *E. coli*, alfalfa, and silkworms. While the scientists are figuring out which system makes the most desirable protein at the cheapest cost, they are also designing applications of the spider silk protein. For the goat system, during the summer months, the lab focuses on purifications, or methods of extracting spider silk from the medium of production, goat milk. For example, the lab is working on a large-scale goat milk method; through a series of filtering and purifying procedures, spider silk is produced in powder form and then reconstituted. I hang out in the lab for many hours during these procedures to observe the processes and interact with the scientists. Spending this type of time in the lab gives me access to emergent innovations in real time.

Early one August morning, postdoctoral fellow Xiaoli Zhang, a scientist who works primarily with transgenic silkworms, enthusiastically greets me at the cubicles. She seems excited and quickly describes her task of the day. I am not prepared for the numerical details she uses to explain her work. At first, I think she is describing a math problem. "We will start with ten strands," she says, "and braid them together to form twelve new strands. Those will become three strands, which we will braid into one strand about seven feet long. And then this will make a rope used to suspend the guy. We will give that to the BBC." When I scrunch up my face, revealing my confusion, she directs me to the lab bench where ten strands of spider silk are taped down resembling the strings on a guitar. Seeming incongruous to this scene is a pink plastic device, the Braid X-press in the hands of an undergraduate student.[43] Using a chopstick, Xiaoli carefully pulls up the end of one strand and, with the student's help, threads the strands through different spokes on the device. Once the Braid X-press is fully threaded, the student presses the button and the device braids the strands together.

Eventually, thinner strands are braided with others, and the final product is a spider silk rope about as thick as a marker pen. This rope will be used to suspend a 150-pound human being high over the Green River in Moab, Utah. As postdoctoral fellow by the name of Thomas Harris explained, this project is a commission from a television program used to demonstrate the strength of spider silk, but since the lab is not an industrial manufacturer or fabricator, the scientists must come up with innovative ways to get the work done. This is not quite how Spider-Man would do it.

Figures 1.1a–c. Images from the Spider Silk Lab: poster on walk-in freezer, blow-up doll of Spider-Man, work cart full of tools. Photos by Lisa Jean Moore.

Figures 1.2a–b. Braid X-press, à la "As Seen on TV," a pink plastic toy for braiding hair; used in the lab to braid spider silk. Photos by Lisa Jean Moore.

The lab overengineered the rope to support 200 pounds and tested it before delivering it to the television crew. Despite its strength, I can't help but wonder how the individual to be suspended would feel seeing this laboratory filled with assembly-by-hand gear and repurposed children's toys. In this moment, the promise of seeing the future of science in action feels so much less boring and so much more playful than I could have imagined. The TV program *Beast Kept Secrets* on Animal Planet, with a tagline "Revealing the truth behind incredible myths surrounding the animal kingdom," used the spider silk on the premiere episode airing in 2021.[44] The two British hosts engage in a variety of strength tests between steel and spider dragline silk; the episode is stylized in the spirit of two blokes who can't believe the ingenuity of animals. A variety of experts weighs in, describing the orb weaver's silk, and the episode culminates as one host dramatically dangles from spider silk three hundred meters above a thirty-meter expanse of the Green River. "It works!" the host shouts. "Spider silk is stronger than steel. There's the proof, man."

These types of side projects commissioned by news media outlets and popular science journalists are familiar to the scientists. Randy notes, "I am sure we have had at least twenty different TV filmings, close to that many radio interviews, and many more than that for print media." Even though these media inquiries (and scholarly requests like my own) are time- and resource-consuming, it occurs to me that Randy has created a laboratory environment that should be responsive to external curiosity. As he says, "I feel we have some obligation to tell about what we do and make sure that it is portrayed accurately, especially with the goats."

The spin of a media event favors the sexiness and quirkiness of sciences elevated and abstracted from the mundanity and hard work. I worry that I might be doing it too, abstracting science and romanticizing it, even now, by talking about fun and toys. On my arrival at the lab, I noticed this tendency. I also realize that science was a lot of tedious, arduous, repetitive, and boring tasks too, and these aspects of it are much harder to write about. The endless testing and retesting that makes up the bulk of science doesn't make for good copy.

South Farm

The actual spider goats live on a large, sprawling collection of barns and sheds collectively called South Farm. Officially called USU Animal Science Farm, South Farm is about an eight-mile drive from the laboratory to the outskirts of Logan. It is a facility of multiple barns, enclosures, and pastures. Starting as a land grant college in 1888, USU takes advantage of the ample land surrounding the university. Programs in veterinary sciences, agriculture, ranching, equestrian management, and equine therapy are just a few of the offerings that make use of cows, sheep, horses, and goats cared for at the facility. The spider goats and nontransgenic goats, bucks, and kids are cared for by herdsmen, USU students Amber Thorton and Andrew Jones, who are responsible for the twice-daily milking, feeding, and tending of the herd.[45]

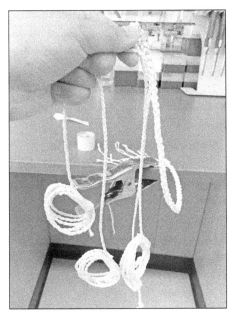
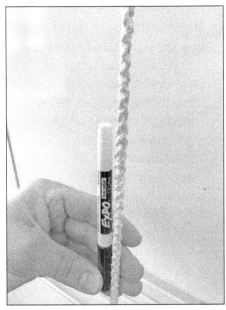

Figures 1.3a-c. Spider Silk Lab fabrication of the rope for use by Animal Planet. Photos by Lisa Jean Moore.

In addition to observing laboratory work, I spent many hours at the farm with the goats and herdsmen (and a pair of mother and daughter barn cats, Cat and Kitty). By mandate of the US Department of Agriculture (USDA), these transgenic goats are quarantined in a secure facility, isolated from nontransgenic species, so that they can never accidentally reproduce and spread the spider gene. Fears of contagion from queer reproducers apply to all species, I guess.

I've traveled to Utah twice over two summers to spend time recording the goats' habits, behaviors, and utilization.[46] Driving onto the farm always seems disorienting because all the unmarked buildings look the same and the GPS doesn't indicate which is the goat barn. Dusty, rocky debris churns in the wake of the truck. The gravely roads are bumpy and pitted. I bounce about in the truck cab, craning my neck as I turn corners wondering where the goats are. Cows poke their heads through the metal fence and chew on fresh hay. They barely regard me as I drive by, even though I can't help but always call out "Moo," an effect of having raised three children. Sheep graze in the pasture. Chewing on my chapped lips, I park behind the last barn. As I step down from the truck, I hear the bleating, plaintive *baas* resembling human baby cries. Goat heads pop up as I step into the barn. The bleating grows more insistent, the smell of hay and goaty odors. The pens on either side of the barn are full of white goats, ears tagged, some spray-painted with a blue streak to indicate microchipping. In the first pen, the two bearded billies slowly make their way to my outstretched arm, and since I'm empty-handed, they turn away. I bend down to get a better look at their very large testicles hanging down, as Amber, the herdsman, greets me holding a long stick. I smile, caught looking at the massive sacs, and stretch out my hand to shake hers.

At the farm, I milk goats, transgenic and nontransgenic alike. I feed them and walk them back to their pens. And I pet them, even as I am sometimes embarrassed by being affectively captured by their cuteness. I work to cultivate intraspecies mindfulness; this practice of speculation about nonhuman species strives to resist anthropomorphic reflections even while acknowledging the empathetic understanding of nonhuman animals (that irresistible cuteness!). It is an attempt to get at, and with, another species to move outside ourselves. I am tracing the threads of who we and spiders and goats are: warm machines (living, breathing machines) and centuries of semiotechnical infrastructures, from histories of agriculture to companion species, from experimental objects to warm machines made war machines. Getting with the animal, or as Gilles Deleuze and Felix Guattari describe, becoming animal, requires new modes of embodied attention and awareness.[47] In my practices with bees and my work with horseshoe crabs, I used my own sensory tools of seeing, hearing, touching, tasting, smelling—their bodies, their habitats, and their products.[48] My analysis of South Farm makes up the bulk of chapter 2.

Interviews

Beyond the interviews I conducted in Utah, I have also interviewed several people by phone and in person. Back in New York City, I take the subway uptown, feeling the familiar jostles and bumps along the tracks. I squish into a seat between two commuters. Utah had taken on the feeling of a dreamy Terence Malick sequence. I'm headed to meet Cheryl Hayashi, PhD and director of comparative biology research at the American Museum of Natural History, a biologist with extensive experience with spiders.[49] The museum, a revered anthropological and natural-scientific institution is an auspicious place to find one of the world's leading spider silk experts. I am greeted by a seated Teddy Roosevelt on a bench in the staff entrance and walk past the Great Hall of the North American Indian, an exhibit that hasn't changed since its creation, being a meta museum object in itself, a token of earlier forms of knowledge about nature and culture. I've been to this museum many times with my kids and spent hours looking at the squid and the whale in their iconic setting for the same-named film about familial queerness in its own feeble bourgeois Brooklyn early 2000s sense. My youngest daughter spent the night here for a "Night at the Museum." My eldest, Grace, has worked here as a summer intern researching paleobotanical DNA in packrat middens.[50]

Cheryl is the start of what's called a snowball sample; she got the ball rolling, and through her, I recruited many others, growing my number of informants. It was through Cheryl that I had the chance to interview several other entomologists, as described in chapter 1, during a site visit to Florida. It's remarkable to me how, over the three years of studying spider goats, the project brought me from Utah to New York to Florida and included interviews with people in Slovenia, California, and Maryland.

Chapter Outlines

My purposes in writing this book are layered. *Our Transgenic Future* documents the invention, existence, and daily life of a herd of spider goats in Utah. I introduce some of the players (human and nonhuman) and how they've created objects, processes, and concepts in this more-than-human world. I explain how the technical, social, and cultural simultaneously produce both the material of spider silk protein and the immaterial of symbols, discourses, and affects. I make claims about how this transgenic creation generates new opportunities for me to comment on emerging science, technology, capitalism, feminism, maternalism, reproduction, ethics, queerness, and animal studies.

Through my fieldwork with arachnologists, herdsmen, and transgenic biotechnologists, I've designed this book to push us to consider the ways humans, goats, and spiders intertwine in the drive to produce what was initially only intended to be military gear and what is now more everyday products like sporting apparel and face creams. In *Our Transgenic Future*, each chapter examines a particular area of research in the making of the transgenic goats.

Because the story of spider goats is a complex one, I have forced myself to slow down as a writer. Crudely, chapters 1 and 2 are primarily about objects—the spiders (silk) and the goats (milk). Chapters 3 and 4 are about processes—purification and product innovation and manufacture. Chapter 5 is about concepts—self-awareness and obsolescence. Of course, this is not a rigid distinction, as objects, processes, and concepts are evident in each chapter.

Chapter 1 begins with the spiders, examining how and why golden orb spiders came to be the spider of choice for the transgenic goats. Humans often have a deep disgust for spiders. As with many invertebrates, humans do not muster the same degree of compassion for spiders when these animals are used in scientific experimentation. Mining the various effects of arachnids, this chapter examines how they are objects of fear and are loaded with symbolic meaning in mythology and in psychology (arachnophobia).

Through fieldwork in Gainesville, Florida, and interviews with expert entomologists, I provide background on the spiders and their wondrous abilities. This chapter also addresses the gendered implications of needing females for reproduction or genes to create the spider silk protein. Billy goats and male spiders are limited to harvesting reproductive raw material, sperm, and are otherwise disposed of by humans or, in the case of spiders, by the female spider herself, when she eats him after insemination.

In chapter 2, I turn to the goats themselves and the cultivation of their lives. I describe their anatomy, general habits, evolution, and domestication by humans. I demonstrate how so many humans have a deep affection for goats, as expressed in popular culture, very much in contrast to how we feel about spiders. Not surprisingly, there are many claims about the general health and well-being of these university spider goats. I also consider goats in culture as stereotypically stubborn and meme-worthy as adorable.

The scientists who raise these goats, milk them, and care for them believe that these animals have a better life than do any other goats. The university goats are cared for in a temperature-controlled enclosure and given ample space. They are nutritiously fed and receive top-notch veterinary care. Members of the public who worry about the continuum of animal cruelty consider the goats prisoners in an abusive cycle of forced reproduction and unnatural living. And regulators, including people in the FDA and USDA, who establish criteria for the maintenance and disposal of the goats, think these creatures are hazardous and must be carefully tracked and controlled to prevent their proliferation in the general population.

Using a critical animal studies perspective, I entertain in chapters 2 and 3 an ontology of the nonhuman—the goats' and the spiders' ontology—to reposition the animals by decentering my own human gaze. In addition to decentering myself, I have also come to see myself differently. Throughout the book, I continue to explore the idea of intraspecies mindfulness, which sociologist and colleague Mary Kosut and I discussed in our book *Buzz: Urban Beekeeping and the Power of the Bee*. As I demonstrate, the quality of life of animals, especially spider goats, is not that easy to determine.

Chapter 3 looks at the goat as a system, introducing some of the key human players in the spider silk transgenic landscape. Meeting with spider goat scientists and others, I explore how humans have co-constructed spider silk as a potential panacea. I introduce the field of synthetic biology as significant at this historical moment. Research in this field and its interpretation will be especially interesting since the work scientists are doing is so sensationalized for most people. This chapter explores the scientists, their language, the machines and techniques they use, and the ways they define objects, materials, processes, and goals. I engage with the more-than-humanness of objects seriously and include nonliving objects. Focusing on the concept of purification, this chapter explores how highly modified goat milk becomes waste in the pursuit of spider silk protein.

In chapter 4, I explore the economics of creating spider silk and provide a meta-analysis of the corporate, pharmaceutical, and military investments and financing of spider silk protein. Spider webs have long fascinated scientists and naturalists for the potential application in biomimetics (also known as biomimicry), or the imitation of nonhuman natural systems or elements for engineering solutions to human problems. Putting aside the significant trouble with potentially privatizing nature as intellectual property, the spider's web has generated many creative scientific endeavors, in particular the remarkable potential of silk protein molecules produced by spiders.[51] Bandage adhesives, biocompatible ligaments and tendons, synthetic muscles, bulletproof vests, and shrapnel-repellent pants are some of the products that have been fabricated or proposed.

The USU Spider Silk Laboratory is divided into teams of researchers who are simultaneously extracting spider silk and innovating products. In this chapter, I follow the different teams as they strive to invent applications of spider silk while they refine the extraction methods. I also introduce the types of spider silk products (some produced from purified goat milk and some from other transgenic models) that have been manufactured and are in stages of research and development. In Randy Lewis's lab, there is a sense of urgency about creating an affordable and marketable product that can be mass-produced. In the case of spider goats, a new species was fabricated to specifically labor for potential economic value that has not yet been realized. I explore what happens to animals when they present a potential to accumulate capital (economic or cultural) and when they are turned into workers.

Chapter 4 concludes a summary of how spider silk transforms the capacities of the military and the medical-industrial complex. Beyond whatever comes next for the spider silk industry and the spider goats themselves, we now have multiple generations of self-producing spider goats. What would it mean if the modified goats went wild? In human history, Africanized honeybees, genetically modified seed, and other invasive species have leaked outside human containment to breed and infiltrate the natural population (such as farmed salmon). Beyond dystopian fiction à la Margaret Atwood, this chapter forecasts some concrete possibilities of this type of slippage between the natural and the synthetic.

In the book's conclusion, I return to the threads that bind all the discussions together. I weave together the book's discussions of goats, spiders, scientists, lab objects, war machines (the whole apparatus of infrastructure, technology, blood capital, and geopolitics) and a dense history of narrative and semiotics. I start with Greek mythology (and chimeras that already throw us back to this semiotic moment) and move on to the microreproductive communities, which include sperm donors and me; multiple forms of parenting and siblinghood and kinship; and how gender and sex inflects, and is inflected, by these forms.

1

Spider Encounters

Silking Comes First

Spiders, you know, honestly, they have some pretty bad PR hurdles to get over.
—Cheryl Hayashi

A Proustian memory strikes whenever I hear Debbie Reynolds's voice. I am automatically transfixed and transported. The feeling of the red plush velour cushion, rocking my feet back and forth to soothe my anxious little body, quickly tapping the sticky black patent leather Mary Janes. The itchy crinoline on my tights as Granny Jean's tiny, wrinkly hands pat my skirt. It crinkles on my lap. My anticipation builds, waiting for the luscious velvety curtains to part. It is 1973. I am six, and my Italian grandmother has taken me to see *Charlotte's Web* at Radio City Music Hall, my first time in such a grand theater. This is where I meet Charlotte. And when she appears, it's love at first sight, the most remarkable creature—"in a class by herself."[1]

Wilbur the pig inquires, "What is your name, please?"

"Charlotte A. Cavatica," she replies, dangling from the barn ceiling.

"I think you're beautiful," he gushes.

And the confident spider admits, "Well, I am pretty. Almost all spiders are nice-looking. I'm not as flashy as some, but I'll do."[2]

Her devotion to Wilbur thrills me as she ingeniously solves his problems, even repeatedly saving his life from slaughter. But when Charlotte dies, I am devastated, and no matter how many times I later ask my mother why, she cannot offer any satisfactory explanation.

Having seen the film several times since and reading the book countless more, I wonder if this formative experience has something to do with my interest in this research project.[3] And because I feel I have this real and even latent connection to Charlotte and maybe even actual orb weaver spiders, it is even more surprising to me that when starting this project, I didn't even register the actual spider as a part of silk milk.[4] Like others I've spoken to, when introduced to the idea of spider goats and their capacity to lactate spider silk protein in their milk, I concocted an image of a goat spraying webs Spider-Man-style from her udders. Furthermore, in discussing transgenic spider silk with friends, family, and acquaintances, there is a deep and ever-present popular concern for the goats. What happens to them in the transgenic creation? How is the alteration in their lactation potentially dangerous for the goats? Were any goats killed in the making of this technology? For most invertebrates, however, humans do not muster the same degree of compassion when these animals without backbones are used in scientific experimentation.

And spiders are not in the insect family. In fact, with my collaborators from Scotland, I examined invertebrate representation in human-animal scholarship and found that despite being more than 95 percent of the animal kingdom, invertebrates are woefully underrepresented in the interdisciplinary field of animal studies. In light of our study, we argue that specific taxonomies of invertebrate animals are being systematically excluded from policy, public, and scholarly debates about animal cruelty and other animal-welfare-related issues.[5]

In fact, no one has ever asked me about the well-being or mortality of spiders in the creation of spider silk protein applications. I myself didn't think much about them at first. Even more striking, before starting this research project, I wrote an entire book about horseshoe crabs and conducted more than three years of fieldwork on the animals. Horseshoe crabs and spiders are closely related.[6] And yet, for months in my early inquiries into this project, and even with my knowledge and respect for arthropods, I privileged the goat and her fantastic capacities for being manipulated.

In addition to horseshoe crabs, I've also written about another invertebrate species, honeybees and their role in extraction economies. Beyond providing their honey for us, honeybees are also forced to pollinate monocrops as part of industrial agriculture, much as horseshoe crabs are bled for their amebocytes to use in pharmaceutical applications, often at grave consequence to the animals. I began my research with no

Figure 1.1. My fantasy Super Spider Goat ejecting her silk. Drawing by C. Ray Borck, 2020.

conscious concern for the spider or the consequences of harvesting its silk. My selective speciesism enables me and other humans to act differently toward different animals.[7]

Add to this speciesism that many humans have a deep disgust for spiders in particular. We even have a name for a phobia dedicated just to spiders. Arachnophobia is the unreasonable and paralyzing fear of spiders and scorpions. Theories abound about why humans are fearful of spiders, ideas ranging from the sociobiological and evolutionary (few spiders are dangerous and can bite you with venom) to the sexology-based masochistic fear and desire of being enveloped in a web of a black widow. The spider web tattoo on the elbow can signify a prison term, a crime, or white supremacy.[8] Reality television shows have capitalized on arachnophobia, where people are offered a million dollars to stay in a tank with spiders crawling all over them for two minutes. This phobia affects between 3.5 and 6.1 percent of the population.[9] Over the years, having witnessed my fair share of students experiencing panic attacks, fainting, crying, or vomiting at the mere suggestion of a spider in the classroom, I am well aware of the power of the specter of the spider. Even the hint of the cure for this debilitating fear—exposure therapy—outrages people who have arachnophobia. Perhaps hope is not lost, as at least one scientific experimental study found that seven seconds of exposure to scenes from superhero films such as Spider-Man can reduce arachnophobia symptoms.[10] If your first instinct reading this chapter is fear or squeamishness, maybe a bit more information about spiders could have the same impact on you as superhero films.

As we live in the geologic period called the Anthropocene, why should we care particularly about spiders, among all the other animals that are facing extinction? How can we not? As an integral part of the food chain, spiders are vital to the planet's ecosystem. As predators of a variety of insects, spiders' hunting maintains homeostasis in various environments, including farmland and intimate spaces of your home. They are also a food source for several species. But humans have a real problem empathizing with the peril of invertebrate animals in these precarious times. It's hard to look a spider in the eye since it typically has between six and eight.

In this chapter, I introduce spiders more generally and the golden orb weaver specifically, describing my interactions with several human experts and a field trip to Florida to gather spiders. Since my epistemology flows from my corporeal and feminist commitment to embodied and situated knowing, I needed to see the spiders, touch them, and watch them in their native habitat. When studying animals, I highly value phenomenological and mythological engagement—holding the horseshoe crabs, performing a beehive check, collecting spiders. I want to be in the world with the spiders (in however limited a way is possible as a human) in ways that are messy, dirty, textured, and visceral. Part of how I get to know them is by sharing our physicality.

My methods for knowing spiders (and all animals) enable me to work against the human normative judgments about the species. I feel I must handle spiders to counter the speciesist media depictions. I want to know these animals, even as I acknowledge that my empiricism (empiricist desires?) situates me on a morally loaded continuum with other humans, from scientists who study them to the arachnophobes who squish them.

Even though I may make a posthumanist argument that my work potentially benefits spiders as a species in the long run, I am contributing to individual spiders' dislocation from their web and abrupt displacement from their habitat. Later in this chapter, I describe my field trip to Florida and elucidate the history of military interest in spider silk and the unclear origins of the golden orb weaver spider's selection for use in creating silk milk.

Spiders: A Brief Overview

Spiders are found everywhere on earth and have been around for more than three hundred million years, according to the fossil record. There are over forty thousand named species of spiders, but it is thought that there are three times as many as yet undiscovered.[11] Among other differences, spiders are distinct from insects in that almost all spiders have eight legs and produce silk. While some other invertebrates, including butterflies, moths, raspy crickets, silverfish, and mayflies, produce silk, spiders make many types of silk that have different mechanical properties and uses. As one of my informants persuasively explained, "I think you could also touch on the stunning ways spiders in general deploy silk to capture prey. Most people don't really appreciate how devious and 'intelligent' some of these spiders are in their silk use. I think spiders should get more credit on this front, and we should really recognize how clever they are. They certainly aren't the dummies or lower lifeforms most people expect they are."[12]

I've come to have a certain reverence for spiders as I conducted research for this book. The golden orb weaver, the species of spider used to make spider goats, makes seven types of silk.[13] Spiders make silk through glands in their abdomens. They spin chains of connected liquid proteins into solid strands by pulling through their spinnerets with their legs.[14] Each type of silk has different strengths and elasticities. Scientists have identified each spinneret used in producing these distinct silks, assessed the functions of each, and determined its amino acid

Figure 1.2. Golden orb weaver spider, *Trichonephila clavipes*, October 12, 2019, Gainesville, Florida. Photo by C. Ray Borck.

composition.[15] For the spiders, the silk serves a different purpose: it is a home, a means of catching prey and collecting water, a site for mating, a protection for the young in egg sacs, a safe retreat from the weather, and a mechanism for ballooning.

Ballooning, also referred to as kiting, is the airborne movement a spider can create by releasing strands of silk from its spinnerets and catching currents of wind to transport itself. The process is described in *Charlotte's Web* when the spiderlings emerge from the egg sac for their virgin aerial voyage: "Then another baby spider crawled to the top of the fence, stood on her head, made a balloon, and sailed away. Then another spider. Then another. The air was soon filled with tiny balloons, each balloon carrying a spider."[16] I wonder if this is where Stan Lee and Steve Ditko in 1962 got the idea for Spider-Man's ability to use webs ejected from his wrists to travel between skyscrapers—a masculine spin on web transportation.

Seventeenth-century British physician, philosopher, and naturalist Martin Lister (1639–1712) is reportedly the first arachnologist and the first conchologist. He is credited with discovering ballooning in spiders. In a letter to his mentor, John Ray, in November 1668, Lister described ballooning in phallic terms:

> It so happened that in a few days, while I was studying the craftsmanship of other Spiders familiar to me, suddenly the one I was watching left off what it was doing, and bending backwards it pointed its anus into the wind and shot forth a thread in exactly the way a strapping young man expels urine from his swollen bladder. I was surprised at the creature's unusual behavior, then I saw the thread stretched for many feet and waved in the air; soon, however, the Spider herself jumped upon it and was swept away wherever the thread took her, while still clinging tightly to it, and it was borne over some quite tall trees.[17]

Remarkably, Lister trained his teenage daughters Anna and Susanna (the Lister sisters) to sketch or engrave images for his scientific publications.[18]

Throughout history, there have been many documented human uses for spider silk.[19] Anthropological accounts describe islanders in New Guinea using spider silk as fishing nets for small fish or rolled to serve as fishing line for larger catches.[20] The use of spider silk in the textile industry appears to have emerged in the 1700s in France with various attempts to procure spider silk from webs and cocoons to manufacture clothing, with poor results of clothing that tore or disintegrated.[21] Before contemporary technological innovation, coming up with a scalable way of making spider silk garments was deeply challenging. It is very time-consuming and labor-intensive to collect enough existing spider silk simply by finding webs. And it is extremely difficult to raise spiders in captivity to produce spider silk. They are often territorial and sometimes cannibalistic, and they require space and care for their survival. As discussed in chapter 3, when scientists, with the advent of genetic engineering techniques, could isolate the spider silk proteins and modify existing biological systems to produce spider silk, spiders themselves were no longer needed to produce this substance.

Golden Orb Weavers

At the ready in the middle of the web, the golden orb weaver spider (*Trichonephila clavipes*) resembles a floating orchid with slender striped legs symmetrically positioned. On top, the body is brownish, yellowish, or olive, and the underside is more burgundy to pink. When I first saw these spiders, I instantly thought of flowers whose petals are so delicate and vibrantly colored. Female *T. clavipes* have orange, yellow, or brown legs with a silvery carapace and are notable for the tufts of black hair on their first, second, and fourth pair of jointed legs. Males are darker brown, much smaller, and not as flamboyant; they are easy to miss as they lurk on the edges of the web.

The silk of this species is a noticeable golden yellow especially when it shimmers in the sunlight, and the web color is where the common name comes from. Orb weavers are so called because of the circular wheel-type webs they create, similar to a spooky Halloween drawing. To understand the female's web is to take in the skillful construction of the spider and, as anthropologist Tim Ingold suggests, to see spiders as nonhuman agents.[22] These webs are made of different types of spider silk and are constructed in about thirty minutes to an hour.[23] The spiders typically position themselves in the center of the sticky web and wait to feel thread vibrations of prey or mates (potentially becoming prey) along the surface of the webs. Their movements up and down the web are graceful, seemingly effortless, and fast. One scientist in Utah insisted that these spiders were "absolutely gentle giants who didn't bite unless extremely agitated."

Golden orb weaver spiders are sometimes referred to as banana spiders and belong to the oldest known surviving genus, which is 165 million years old.[24] The taxon was first described by Linnaeus in 1767 and was originally called *Nephila clavipes*, a name derived from ancient Greek and meaning "fond of spinning."[25] Golden orb weavers are part of the Araneidae (orb weaver) family, which makes up one-quarter of all spider species.[26]

In 2019 a team of arachnologists led by Slovenian arachnologist Matjaz Kuntner changed the genus name from *Nephila* to *Trichonephila* (*tricho* in ancient Greek means "hairiness").[27] Over a Skype call, when I asked Matjaz why this change, he replied matter-of-factly, "Of course, we do this all the time as taxonomists, where new evidence comes to light about phylogenetic differences and we make nomenclature classification changes. *Nephila* are a well-known genus, but they violate a rule of monophylogeny, so they need to be split in two genera."

I smiled because I was on the very edge of comprehension. "So basically, they didn't fit into a category," I said.

He nodded. "Yes, they did not."

Beginning to learn about these spiders, I am met with the basic facts. For example, these spiders are pantropical and, in the United States, are habitually found in the southeastern region, specifically Florida. Their life span is about a year. They have eight eyes but also use their legs to sense and locate prey, and their bodies are divided into two parts. The cephalothorax, the head part, is where all eight legs attach, and the abdomen, the larger of the two parts, contains the silk glands, the genitals, and the spinnerets. Spiders must maintain their webs and fix or rebuild damaged parts frequently, depending on wind and rain and capture. When fixing their broken web, spiders eat the fragmented strands and use them to generate fresh web for repairs.

Beyond these and other facts about golden orb weavers, I was unsure of what specifically drew humans to select this spider as "the one." During our conversation, I asked Matjaz why he thought biological engineers chose *Trichonephila* for transgenic manipulation.

He paused for a second, thinking, and said, "Well, these webs are huge. When you see them, it is very difficult to ignore them. They look strong and feel strong, especially when you run into them. I think it was assumed that the silk was exceptional, but actually it is not superstrong compared to other silk. It is nonexceptional. But there is a great deal of volume."

Living in New York City, I don't see any *T. clavipes* in the wild. To better understand these spiders, in early August 2019 I met with Cheryl Hayashi, the brilliant and generous director of comparative biology research at the American Museum of Natural History. Cheryl had been a postdoctoral fellow in Randy Lewis's lab earlier in her career, when the spider silk work began. Long before she became the MacArthur grant–winning scientist of today, she was, on leaving her home in Hawaii, an undergraduate biology research assistant whose "job was taking care of a professor's lab colony, which happened to be spiders."

Cheryl explained, "I had to do the work of putting crickets and flies in spider webs. And those spiders weren't in cages like this"—she pointed to the cages that hold spiders in her lab. "Instead, they were in an environmental chamber. And so you go inside."

Startled by my mind's image of her petite self walking into a spider chamber, I asked her, "And you had no arachnophobia before that?"

"No, I didn't," she said. "I was actually more afraid of their food jumping around."

Cheryl, like Matjaz, described how impressive the size of the webs can be: three feet in diameter with anchor strands two or three yards long. These spiders catch insects, typically mosquitoes, flies, moths, bees, wasps, beetles, butterflies, or grasshoppers, in their nets and bite to subdue. Their venom is not harmful to humans, although it does have a subduing effect on insects. The golden orb weaver's predators are birds and certain wasps.

Cheryl explained how *Trichonephila*'s dragline silk became the most interesting to scientists: "If you're going to work on one silk, that would be the one to work on, because it's very, very tough—has a moderate, you know, 10 to 15 percent extensibility that makes it have very good mechanical properties."

Later I learned that dragline silk has been the most commonly studied silk after scientists identified the silk's protein. Justin Jones, a biologist and the current director of the Spider Silk Laboratory at USU, described this silk in further detail: "An orb weaver's dragline silk is comprised of two proteins, right? It's the MaSp 1 [major spider ampullate protein 1] and MaSp 2 proteins. OK, so that's why we have the two herds of goats, about four or five animals. One of them produces one of those proteins, and the other produces the other."

During my museum interview, Cheryl continued describing spider silk properties: "Most spider silks are very hydrophobic. That means water just beads right off of it. And they're also inherently antimicrobial. So if something was lined with silk, you might think the water wouldn't settle there." I nodded, unsure of the significance of this fact. Sensing my uncertainty, Cheryl continued, "And there wouldn't be a lot of bacteria. And it is very tough so it won't wear out—it would be very durable, so maybe coatings would last longer. We also don't react to spider silk."

I wondered aloud how that was discovered.

"They implant a little piece of silk or spray it on [the human skin]," she explained, "and there is no reaction. And I think the earliest studies were done with mice."

Spiders, like horseshoe crabs and some honeybees, exhibit sexual size dimorphism. In fact, the size of female spiders is commented on a great deal in the literature. The males are tiny in comparison to females and are mostly black and brown, fairly inconspicuous. Males are about one-quarter inch long, while females can be up to three inches long—not including her legs, each of which can reach five inches. The web of a female may have multiple small males temporarily living in it.

When I brought up this point with Cheryl, she suggested that the size of females has led them to be used in ecological and biological research. "So the female is much larger." She spread out all her fingers. "This does create an interesting thing where, you know, almost everything we know about spider silk is based from studies of female spiders." Large female bodies make for good research subjects in the case of spiders, horseshoe crabs, and honeybees. These three animals are reproductive powerhouses, producing hundreds, even thousands, of offspring in their lifetime.

Much of my intellectual life has been to unpack the sociological scaffolding and psychological naturalization of heterosexuality. For more than two decades, I've taught and written and thought about how heterosexuality obtains material and symbolic assistance to achieve cultural dominance. These processes of "helping out" heterosexuality are then erased and presented as if heterosexuality is just *natural*. Heterosexuality is privileged because the very gender binary on which it rests is an artifact of white supremacist racial hierarchies.

At nearly six feet tall, I am a large human woman, and as such, I am captivated by the sexual size dimorphism of other species where the females are significantly larger than males. Women go to great lengths to appear physically (and presumably performatively) diminutive to men in the expression of legitimate and culturally revered human femininity.[28] Hegemonically, women are trained to believe this and shrink themselves to meet social expectations. When attempting to reinforce an existing human power structure as natural and panhistorical, popular accounts will point to animal behaviors as manifestations of some universal law. For example, *Reader's Digest* and *National Geographic* articles valorize long-term heterosexual monogamy by pointing out animals that mate for

Figure 1.3. The author observes *Trichonephila clavipes* in Gainesville, Florida, as it climbs up its dragline. Photo by C. Ray Borck.

life, including swans, gray wolves, and macaroni penguins.[29] The descriptions, rife with celebratory value judgments about cultural preferences for heterosexuality and monogamy, highlight animals' relationships as "mirror[ing]' those of humans." But humans are selective when it comes to searching for evidence of "natural" behaviors in the animal kingdom. Most mating behaviors are narrated as familiar heterosexual romances.[30]

My own experiences of performing heterosexuality have always felt fraudulent because they are. I tower over my male partners. When holding hands with boyfriends, I know we look weird to some people. There have been many dances and proms in flats, photographs with bent knees, and creative uses of stairs for kisses. What would it be like if we were to look to other species as some referent of the animal kingdom? I don't necessarily want to give more rules for heterosexuality but rather prefer to explode it into all its potential expressions. Why can't females, physically larger, and males, diminutive, be the ultimate expression of peak reproductive achievement? What if these female spiders, horseshoe crabs, and queen bees were my role models? So juicy, big, and sexy, towering over demure males—the ultimate trample-fetish.

I had to see these spiders for myself. In early October 2019, I began contacting some entomologists at the University of Florida. I cold-called and emailed a few scientists, hoping to connect with someone who could introduce me to these spiders in the wild. My efforts paid off, and with a lot of luck, quick planning, and some last minute plane tickets to Florida, I am accompanying a scientist on a collection trip where he and his crew gather specimens of golden orb weavers.

Spider Specimen Collection: Gainesville, Florida

I contact Lawrence "Lary" Reeves, an entomologist at the Florida Medical Entomology Laboratory in Vero Beach, Florida, in mid-October 2019.[31] Although Lary has expertise in mosquitoes and vector-borne diseases, he also collects insects for scientists and museums. Coincidentally, he is about to head to Gainesville to collect seventy-five golden orb weaver spiders for an exhibition at the Natural History Museum of Los Angeles County. For the month of October, the museum transforms its butterfly tent into a Halloween experience for visitors; spiders and their webs are the main stars. Lary has generously allowed me to accompany him on his collection field trip.

Before the expedition, Lary advised me to wear long pants and closed-toe shoes, explaining in an email:

> This time of year there will be a lot of plants that produce hitchhiking seeds e.g., Bidens alba, sand spurs, Desmodium. You can expect that pants will be covered in seeds! There are also ticks and chiggers in these areas (fortunately, in Florida Lyme disease is not the issue that it is up in the Northeast). I don't use insect repellent very often, but it can help to avoid ticks and chiggers. It should still be pretty warm in Gainesville this time of year, so in general, I wear long pants and short sleeves for fieldwork. I'd keep a light jacket handy in case the mosquitoes are bad, but they should not be this time of year.

I prepared for the trip to Gainesville, checking in with a friend. "I can't come to get drinks," I told her, "because I'm going to be in Florida to collect spider specimens with this entomologist."

She gasped. "Please be careful," she said, "because it's creepy and weird and sounds like a horror movie."

I considered myself forewarned, but any nervousness evaporates on my meeting Lary, a friendly, kind-faced scientist with visible tattoos, including the double-helix DNA and multiple flowers. I spot in his Subaru a Spider-Man car seat, which he later says his four-year-old daughter insisted on.

I am surprised by the lusciousness of vegetation in west Gainesville. Lary is familiar with the area ("I spent many hours as a child riding my bike around here with my butterfly net"), and he knew of several locations where *T. clavipes* proliferates. We drive around single-lane county roads, stopping at nondescript locations in a wooded area and walking along the brush line. These spaces at first feel like interstices, or forgotten spaces seemingly devoid of function, rather than teeming with life, a rich habitat for nonhuman animals.[32] Perhaps the very quality of being a no-man's-land of space creates a protection of the habitat.

Pulling the rental up the embankment at the side of the road, I hop out of the car. We stand and stare straight ahead into the woods. I see nothing but tangled branches and layers of green leaves. I'm ready to pack it in and try the next spot. But Lary tells me to look straight at eye level and adjust my focus to places between the vegetation to where sunlight pours through. The spiders set their webs up to capture as much insect traffic as possible in the clear spaces through the trees. "You can see the color, a little golden in the light of a web. And you will see a female in the middle of the web; they are getting bigger this time of year, getting older. They've finished molting and are getting ready to spend a lot of energy laying eggs." I strain my eyes and become frustrated, still seeing nothing. "The egg sacs will overwinter in the fall and the babies will get ready to hatch in the spring," he continues. "They are sputtering out now, getting ready to make their egg sacs. Then the adults will die when it gets colder."

I flinch at this term, *sputtering out*, in an (over)identification with all female animals. I want to ask Lary, "So, am I sputtering out here in menopause?" but I remain silent, realizing this could be read as hostility, or worse, menopausal rage.

I work to contain my impatience and worry that this entire trip was a preposterous idea, thinking I could fly down to Florida for twenty-four hours and actually find the spiders.

"Here's one," Lary says. He twists a dead branch into a web and pulls it down to our waist level. At the end of the branch is a spider.

I gasp and my eyes surprisingly tear up. "She is beautiful," I say. The morning light is highlighting her from behind, and she looks like a jewel with a holy glow. I am astounded, bewitched by her artistry. She is motionless, and I learn that since these spiders are mostly nocturnal, the daytime is often a time of stillness. Centered stillness.

Lary beams at me, pleased that I am so delighted. "It's so great to find someone who likes them as much as I do!" he gushes.

I realize I am holding my breath not from fear but from wonder and delight. The spider crawls along my finger several times and around my hand and arm. It tickles my skin, and I roll my wrists around as she goes from front to back and over my fingertips. It feels like she has sticky single toes on the ends of her legs. Grabbing on to some nearly invisible silk attached to her body, I lift my hand up and watch as she dangles at the end of the line. I'm amazed as she climbs back up to my hand using leg over leg to reach the top.

As I observe this spider, her long, tapered legs operate separately, each reaching for a different point of contact and moving quickly to secure a position for her entire body at the top of my hand. She seems to be exploring me, but I imagine she's also anxious to get back to her web so she can make her egg sacs or catch a meal. I wonder if these spiders know they are about to be mothers. Is *mother* even the term? Do they feel an urge to make the egg sacs that will pass on their genetic material? Do they protect the egg sacs until they die? Since they will die well before their offspring have emerged, do they connect what they are doing with their spiderlings? What is it like to be a mother who never sees her kin? I pull my hand closer to my face so I can attempt to look into her eyes. But just then, Lary then plucks her off my palm and places her in a plastic container with a small bit of paper towel at the bottom.

She is now trapped inside this new sterile environment, no longer in the fresh air and morning sunlight. I wonder if she had already laid her egg sac and was ready to sputter out. Did she want to check on it a few more times and make sure it was secure? Will she survive until Los

Angeles to perform for humans and then die? Looking at her through the plastic container, I consider how she is both an entertainer and a part of a transgenic invention. She has been made useful to human commerce, both the carnivalesque and the biotechnical.

Throughout the day, Lary points out the metallic red and green dung beetles entangled in the webs, and the kleptoparasitic spiders that hang out on the golden orb weavers' webs, waiting to steal their subdued wrapped prey. He also teaches me how to identify when a web is empty by judging the level of debris hanging in the dwindling mesh. I feel more confident as the day progresses and start to spot webs that dramatically decorate the in-between spaces of forest. These spiders are expert hunters and are able to build their webs in the flight path of insects. Even though males are harder to see because of their size, coloring, and location, once I am attuned to the spiders, I can identify the males at the edges of the web. Male orb weavers rely on females to make webs and catch prey—living on the outskirts of the web their entire lives. Males also steer clear of females other than to mate, and even when mating, they approach her carefully, as females do sometimes eat males. Having a much less impressive appearance and unable to weave these impressive webs, the males are safe from our collecting expedition. Their inability to spin has also saved them from transgenic dissection.

I continue to romp around in the bushes, turned on by the tactile nature of the work. By touching the spiders, I learn about their size (they span my entire palm), their gentleness as they walk all over me with the most delicate of steps, and their quickness as they are always in motion once off the web, never stopping to rest in my hand. It becomes obvious that they are out of place and seem to want to return to their home.

We travel from one side of the road to another, moving from spot to spot to collect spiders along about two hundred meters of roadway. I repeatedly watch Lary cup spiders in his hands and drop them into plastic containers, stacking them on the ground and transferring them to his car. I call him over when I see a particularly large female or wonder if one is eating something and want Lary's clarification. My clothes are sticky with burrs, my arms marked with bramble scratches and getting sweatier as the morning sun heats up the asphalt. We chat about the different angles of webs and possible hunting strategies and admire the webs out of our reach high in the tree line. At the end of the three hours, my cheeks ache from smiling so much. Shamelessly, I also take some selfies with the female spiders and text them to my daughters.

Lary will collect seventy-five of these spiders to express-ship them to Los Angeles for the natural history museum's Halloween display. They fill the passenger-side floor of his car in plastic containers with holes poked through. They will be shipped in a large box for a next-day delivery. As we cover the area, he describes other times he has gathered spiders. "I don't feel bad, because these spiders have laid their eggs and are at the end of their life cycle. The first substantial cold weather is going to knock them out anyway. And I am all for venues where people see animals. People need to learn to care about these things." Their life cycle is their reproductive cycle, just like Charlotte's; the childhood mystery is solved. As explored in the next chapter, using my maternal experiences as a vantage point, I have found it easier to establish attachments with goats and develop a type of intraspecies mindfulness about them than I have with spiders. Since golden orb weavers are not mammals and they bear offspring so differently, dying before their offspring even exist, I am limited in my ability to imagine the female's connection to her children. The affective and empathetic skills I have tried to cultivate when encountering and interpreting goats do not translate as easily to spiders. And yet I am left to consider how humans' domination of spiders, at least in this collection expedition, dislocated and deprived female spiders of their typical lived experience.

Like other animal studies scholars, I'm not sure it is morally defensible to create so-called animal encounters (such as spider or butterfly rooms, zoos, and dolphin shows) on the grounds that the interactions raise awareness about animals, in this case, spiders.[33] However, creating kinship with nonhuman animals is important for conservation efforts and biodiversity. Introducing humans to nonhumans in educational settings can create a lifelong relationality. Sociologist Linda Kalof has contributed to our understanding of how encountering animals even through photography can create affective bonds from humans toward nonhuman animals.[34] In one collaborative study, Kalof, sociologist Cameron Whitley, and photographer Tim Flach found that viewing close-up animal photographic portraiture is positively associated with increases in human empathy toward animals.[35] These photos, however, were mostly of charismatic megafauna and not spiders. Although

Figure 1.4. Bottom of Lary's car, filled with more than forty spiders. Photo by Lisa Jean Moore.

Charlotte's Web might create an early childhood experience for some, we have little other opportunities for positive arachnid reflection to fortify our cultural consciousness about spiders. But as I look at the pile of plastic-encased spiders in Lary's car, I do fret about the individual spiders' discomfort and dislocation for the broader cause.

Furthermore, I am not convinced we can empirically measure any relationship between animal suffering (e.g., spider dislocation and stress and death), human exposure to animals, and humans' attitude or behavior regarding wildlife conservation. In other words, can we measure the individual spiders' distress? Can this measure then be in a range of acceptability as justification for the positive consciousness-raising of humans toward more compassionate coexistence with the species of spiders? How many individual spiders must go through apparent torture in the hope that the future is brighter for all spiders? Do the ends justify the means?

Something about this process of collecting spiders reminded me about my own youthful work to encourage compassion. As an undergraduate student doing women's collective activism in the late 1980s, I visited high schools and spoke on a panel as "the bisexual" among other queer friends. The panels were part of an awareness-raising campaign to hopefully diminish homophobia in suburban Boston high schools. Sociopsychological studies have shown that contact with a LGBTQ person can shift (presumably heterosexual) individuals' negative attitudes about gay and lesbian people, stigma, and gay rights.[36] These findings support the *contact hypothesis*, which suggests a measurable reduction in prejudice through intergroup contact.[37] After a couple of semesters doing these panels, I stopped because I felt increasingly uncomfortable with the "ask any question you want; there is nothing off-limits" free-for-all that often happened.

Obviously, I consented to participate in these events, unlike the spiders, and could extricate myself (despite some very real peer pressure). I was not captured and put into a container to be shipped somewhere for the alleged benefit of my fellow queer comrades. But as I compare these experiences, I see similarities. Both experiences privilege a dominant group, humans or school administrators, basing an activity on the exposure of those more vulnerable, claiming that contact would achieve something deemed good on a meta level for the vulnerable. The collateral damage to those who are asked to perform was considered worth the benefit. Fewer queer kids bullied, fewer spiders exterminated?

Trichonephila Reproduction

Later that day, I conduct an interview with G. B. Edwards, an entomologist, a museum curator, and a taxonomist who for thirty-seven "and a half" years worked for the Florida Department of Agriculture identifying spiders (there are more than nine hundred known species in the state). Several of the University of Florida entomologists have suggested I speak with him. He shares that he "became, basically, the Florida Spider-Man."[38] While texting to arrange our meeting place, true to his work title, he wrote, "I'll be wearing a black Spider-Man T-shirt." I am immediately smitten by the fanboy flair.

As a folk musician plays bad covers of 1980s songs in an outdoor pavilion, our conversation is lively and engaged. G. B. lights up when he speaks of the best part of his job. "We try to combat arachnophobia through education—in fact, one of the biggest things I enjoyed about my job was being at a fair and getting someone who is deathly afraid of spiders to hold a tarantula." I think of my early-morning collection expedition and wonder if there will be any fewer arachnophobes in Los Angeles, a rationalization I'd like to accept.

Even though I have read about *T. clavipes* reproduction, I ask G. B. to explain it to me.

His description, delivered calmly and thoughtfully, is riveting. "Males do not make their own webs as adults. As they grow up they do, because they have to catch food. But as adults, they just come and hang out and might steal a little food from the female, but they are just on there mostly interested in mating."

I nod and think about the males I saw today just waiting around, eating food they did not catch. (I have a chip on my shoulder about women's work.) During my work with honeybees, many urban beekeepers told me that drones (male bees) "only eat, sleep, and fuck." Arlie Hochschild's work on the second shift immediately comes to mind, and I jot a reminder in my notebook.[39] In 1989, when Hochschild's book came hot off the presses in my senior year of college, I devoured it because it validated my experiences of witnessing heterosexual women working full-time and then returning home to a majority of the housework and childcare. Do certain male invertebrates rely on the overfunctioning of the female of the species? I am self-conscious of my feminist politics' encouraging my anthropomorphism and unexamined humanism. But I am also rankled by the apparently overwhelming evidence of female powerhouse capacities (honeybees, horseshoe crabs, goats, humans) to socially and biologically make and remake species while part of society denigrates them as propping up the scant contributions of men.

G. B. continues: "Once she has done her final molt, typically when she is receptive or molted to maturity, she gives off a pheromone, and he'll pick it up. To gauge her reaction, he will pluck the web because he doesn't want to turn into a prey item. Obviously, the males are way smaller than the females. Which in this case is not thought to be dwarfism on the males' part but [is considered] gigantism on the females' part."

I can hardly hide my offense at his use of this terminology. Being seriously gigantic myself, I scrunch up my face. "Why?" I ask him.

G. B. is gentle in his response. "Because most of their closest relatives are more in the size range of the males. The females can produce a lot more eggs; they can catch bigger prey. There are good reasons to evolve to a larger size." Gigantism seems loaded with pejorative meaning when applied to a human female just as dwarfism does to a human male—but for spiders we can see it as good. G. B. continues:

> The male takes his pedipalp, which means foot hand. It's an appendage that's between the mouth parts and the legs, looks like a small leg. But in adult males, the last two segments are modified to create an almost hypodermic-type situation. So, spiders basically have what could be classified as a natural form of artificial insemination because both sexes have their gonopores on the underside of their abdomen. The males have no primary intermittent organ. So they make what's called a sperm web. It's a very small, densely woven web, and they deposit the semen directly on the web. So they literally suck it up into the palp and carry it around.

He presses each of his own fingertips together into points in each hand, holding his hands up in front of him, mimicking the male spiders. He describes what happens next:

> And then, of course, they're actually—I forget how many, maybe seven or eight different positions that have been classified in spiders—but basically, when the female's hanging, she's hanging head down. So the male comes over and he's plucking around, making sure she's not going to come after him, and then he comes up, and he gets directly venter to venter [underside to underside]. And he just reaches, and it happens that fast.

G. B. pokes his finger into his other hand quickly a few times to demonstrate.

> I mean, it's just, he's just inserting . . . There's a hollow tube, and she has paired gonopores there. He's sticking his embolus in. There's a reservoir inside the tip of the palp with hydraulic pressure because they have an open circulatory system. He squeezes; that forces the sperm out into her. So it happens really quickly.

She stores the sperm until she's ready to lay eggs, and then multiple males can mate with her. In their systems, I am pretty sure it is first in, first out. In spiders, there's two types of systems, and it's either first in, first out, or first in, last out. That means whoever is first in with sperm is first out with offspring. So usually the first in, last out, there is mate-guarding behaviors. That's not *Nephila*—they are first in, first out.

He takes a bite of some chips and nods. He is matter-of-fact, as if he just described a technical maneuver like parking a car. But I sit, literally with my mouth open in awe. Making sure you don't get eaten while you use a foot full of sperm to insert into a gonopore of a giant spider sounds intense. But there is also something deeply empowering to me about a female that has all the sperm she cares to collect and, when she is ready, uses it at her discretion. I remember the liquid nitrogen tanks full of frozen semen. For two months, I signed for them from UPS and then stored them in my living room until I ovulated and drove them across Brooklyn for intrauterine insemination at my doctor's office. More spidery than I realized.

Military Research on Spider Silk

Having walked through webs and handled many spiders as they dangled on their draglines, I now have firsthand knowledge of the strength of their silk. Boarding the airplane back to JFK, I look down at the sticky residue of web attached to my pants. All this female spider labor, an annoyance and frustration to me as I move through transit. The stuff is hard to pull off; it has staying power. I can imagine how people might dream up applications in textiles.

Since at least the 1960s, spider silk has been a preoccupation of the US military. I tracked down a 1968 army report that is cited by many spider silk scientists as one of the foundational documents praising the potentially revolutionary biomechanical properties of spider silk.[40] Randy Lewis described his own initial funding to establish the Spider Silk Laboratory as crucial. "That army report was the basis for convincing the Office of Naval Research to fund our first research grant on spider silk," he said. The US military has been the consistent funder of spider silk research in Randy's laboratory and Cheryl Hayashi's as well. Grantees must argue how their research "might potentially be beneficial or of interest to the Department of Defense; they give a very long list of things they're interested in, and you make a justification of your research in those parameters," Cheryl explained.

The original report describes spider silk's properties of interest: "These strengths class these fibers among the strongest organic fibers known. Equally remarkable, however, are the elongations before rupture—amounting to 15 percent and more." It is difficult to manufacture a fiber that is both strong and elastic, and the military has been interested in these properties in the same substance:

> The importance of the study lies in the confirmation of the existence of very high tenacity, natural polypeptide fibers complemented by relatively large elongations to break. Equally important is the fact that the principal amino acids making up the polypeptides are the relatively simple glycine, alanine, glutamic acid, and proline. This points to the future possibility of synthesizing a polypeptide fiber with properties derived for a specific application.[41]

In 1967, the army scientists went on expeditions to Brazil and Florida to collect spiders, which were then "silked under controlled circumstances" in a lab, to obtain about three thousand feet of spider silk. Five species of spiders were used—*Nephila clavipes* (now named *Trichonephila clavipes*), *Argiope aurantia*, *N. cruentata*, *Parawixla audax*, and *A. argentata*. *Nephila* silk was the strongest of all five species' silk tested. The silks were then analyzed for their amino acid content. The report describes how the spiders were silked:

> The spider is held tightly between thumb and forefinger with its spinneret pointing away from the operator and aligned with the edge of the winding spool. The distance between the spider spinneret and the winding spool is about three inches. The thread is started by applying the sticky surface of a small piece of adhesive tape to the spinneret of the spider, holding it in place, for about 10 seconds and then pulling gently away from the spider. By using this procedure, it was found possible to pull away a continuous strand of spider silk. Induced extrusion of the silk thread material could be continued until the spinning fluids were exhausted or the spider broke

the thread. The thread was taped to one end of the winding spool and the spool driven at constant surface speed by means of a variable speed electric motor. Approximately 20–30 feet of spider silk were taken from each spider. In certain instances spiders were silked until the thread line broke due to lack of fiber forming material within the spider's spinneret.[42]

This late-1960s scene of army scientists silking spiders feels almost comical, like a Monty Python skit. Men with crew cuts, in crisp white short-sleeved, button-down shirts under lab coats crouching over tiny spiders on a lab bench trying to pull out nearly invisible threads. The level of tinkering on these tiny creatures astonishes me, its manual simplicity. I asked G. B. about human silking of spiders and the potential harm to the animal. He responded, "No, silking does not harm them. Might make them hungry, though! After all, you are removing some of their resources, largely made from protein." Cheryl had a similar answer: "In my experience, the spiders that get silked do not appear obviously affected by the procedure. I'm assuming that the spiders are not kept under the microscope for an excessive amount of time."

Cheryl's assumption feels like an odd combination of touching and presumptuous. It is commonplace for humans to speculate on the unknowable subjectivity of animal others; the process requires both empathy and paternalism. Even more so, when we care about other animals, we want to believe that our actions are not harming them or that we do less harm by making modifications that we imagine will soothe them. We even have laws and agencies to protect certain animals from human-induced harm.

The US Animal Welfare Act of 1966 (and amended many times over the years) sets and regulates the standards for humane care of *some* animals sold as pets, exhibited for entertainment, used in research, and transported commercially. Even with the act, there is a subjective caveat that laboratory research on animals should limit their pain "whenever possible" and allows for permission to withhold painkillers.[43] More significantly for spiders, invertebrates are not included in the act. Unlike the humans who work with spider goats, my human spider-expert informants do not describe regulatory visits or bureaucratic hassles for conducting research on or transporting spiders. During the early 1990s, ethnographer M. T. Phillips conducted a three-year study of animal research laboratories and found that these laboratories did not have anesthesiologists whose responsibility it was to administer or monitor anesthesia. There was also limited use of analgesics after surgical experimentation. Most importantly, Phillips contends, "scientists rarely saw any pain or suffering in their labs. Their view of lab animals as statistical aggregates overshadowed any perception of an individual animal's feelings at any given moment."[44] In this case, pain is in the eye of the human scientist. Lab work with animals is a morally messy enactment that requires lab workers and scientists to negotiate –morality.[45]

In 1990 Steven Lombardi, an army molecular biologist with an interest in engineering bulletproof vests and helmets, worked with manipulating bacteria to produce spider silk protein. He suggested that "an all-spider silk vest that could withstand the same bullet or shrapnel impact as the Kevlar vest would be about 30% lighter, cutting a six-pound vest to four pounds . . . the spider silk fiber can elongate or stretch 18%, compared with 4% for Kevlar, and thus can withstand a greater impact without breaking."[46] The stronger-than-Kevlar motto was born, and refinements of extraction and silk expression continued. A later 1993 report about the breakthroughs in genetic engineering described how techniques are being applied to spider silk.[47] This idea of spider silk being an alternative to Kevlar, though perhaps not the chief desire of all scientists, is exploited to maintain the public and the military's attention.

It is difficult to piece together a singular and precise reason why *T. clavipes* was selected. Cheryl said that initially, in the late 1980s, researchers focused on *N. clavipes* (the name back then) "because the dragline was characterized as being so wonderfully strong, it's wonderfully tough, but there was also a practical reason in that it's a very large spider. So in terms of isolating the particular gland and then doing the molecular biology to clone the first spider silk gene, it was easier to do." As previously noted, the webs themselves are expansive, magnificent, golden, and dramatic. Justin Jones, from the Spider Silk Lab at USU, agreed. "I would say this is the precise reason why *N. clavipes* was selected," he said. "The fact that their web silks had been studied and were known to have impressive mechanical ability, combined with their physically impressive size, made them ideal candidates."

Cheryl spoke about the early transgenic research:

> The paper that reported the first [gene] fragment of the first spider silk team was published in 1990 by Randy Lewis and his group. So that was the first time the gene, the spider silk gene, had ever been visualized. Randy's lab cloned the first spider silk gene fragment into *E. coli*; it was a hundred percent pure spider silk with the help of bacteria. The *E. coli* can make fragments of spider silk, but to actually make spider silk a sub protein, [a single] protein molecule that approximates the natural silk molecule—it's very hard because there's nothing, it's just beyond the size of what *E. coli* is evolved to do.[48]

Over time, I often heard that *E. coli* was not a good system for making spider silk protein, because the protein is too long. However, in general, *E. coli* is a model organism or system for science because the microorganism is easy to cultivate and most strains are not pathogenic. Once you get a stock of *E. coli*, you can grow it on agar plates (plastic dishes containing a nutrient gel) to create single colonies that are monoclonal (meaning all the same). The bacteria can be stored, and you can replenish your own stock—therefore it is relatively cheap. But the gene that encodes the spider silk protein is quite long, and *E. coli* is small (about one-tenth the size of most plant or animal cells). There is more space in the animal cell for genetic information. In *E. coli*, there is no nucleus or other organelles; the genetic information is located in the cell. So it is harder for *E. coli* to encode longer genes that require more space.

When I returned home from Florida, I wrote to Randy, asking him about his handling of spiders. I asked if he ever had to engage with the actual spider or if this bench science was already established. He replied almost immediately:

> We silked an innumerable number of spiders of several different species! At one point we were doing about 20 spiders every other day to get enough silk for both physical (mechanical) and biochemical testing. We were the first to clone the genes for each of the different silk proteins made by the orb weavers as well as proteins comprising the major ampullate silk from a wide variety of species throughout the spider family tree.

Despite Randy's unfailing patience for my neophyte and presumably rudimentary questions, I was self-conscious about reaching out to him. That said, I was still confused about the specific use of *T. clavipes*, so I asked him why they selected this spider. Again he responded immediately: "The Golden Orb has been used for more studies than any other spider, it was large enough for easy dissection to get the different glands to use to clone the genes for each different silk, and was available for most of the year. Those were our bases for the choice. We also used a local spider a lot, *Araneus gemmoides* (Cat Face Spider)."

When thinking about all that has gone into making transgenic spider goats, I imagine human innovations for silking spiders and milking goats: placing female animals in human-engineered ("man-made") contraptions that take their natural resources from them. We take the silk from female spiders (and, as we will see in the next chapter, the milk from female goats). Since one angle of Cheryl's scientific research is to characterize the diversity of spider silk, I asked her to explain her procedure for silking spiders when I visited her at the natural history museum lab. She first listed the materials needed to silk spiders:

- carbon dioxide tank to anesthetize the spider, which, once anesthetized, can be positioned and safely restrained onto the microscope stage with transparent tape
- stereomicroscope
- variable-speed rotating motor
- plastic spools

Cheryl then removed a spider from its cage. Using carbon dioxide, she anesthetized the spider and taped it down to a slide. She then placed the spider under a microscope to see which spinneret she wanted to take the silk from, and she pulled the silk from the spinneret. She wound the silk onto a plastic spool, and then, when she was done ("they live to silk another day"), Cheryl placed the spider back into the enclosure.

Like the strands of the web reaching into different trees, these spider silk scientists and innovators are connected to one another. I asked Cheryl if there is anyone else I should check in with about spider silk, and she suggested David Breslauer from Bolt Threads. "I taught him

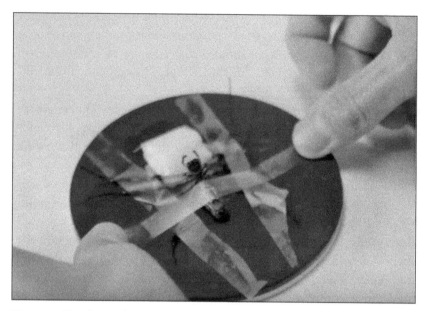

Figure 1.5. Cheryl Hayashi taping down a spider for silking from the spinneret. Used with permission from Luke Groskin, *Science Friday*.

how to dissect spiders," she said. "He was at UC Berkeley, and he flew down to where I was at UC Riverside at the time. He flew down, and my lab group and I showed him how to dissect a spider. For his dissertation, he was trying to make a microfluidic device that would mimic a major ampullate silk gland that he was doing very detailed studies of all the angles. And that kind of raises interest in my spider silk. So he learned how to dissect." I interview David in chapter 4.

Ending my interview with Cheryl at the museum, we were saying our goodbyes when she stopped and turned to me. "You know, I'm kind of thinking about what you're saying about these goats."

I felt relieved that my project had made some impression on Cheryl. Doing qualitative sociology across disciplines, I always fear that the work I do may be unintelligible to natural scientists. I nodded and said, "And I think it's absolutely true that all of us who work on spider silk, even if you're not currently in Randy's lab right now, we've all had the benefit of the goats. It's gotten the lab a lot of attention. And it's in a lot of people's minds. I'm sure it's in a lot of funding of program officers.

You know, once you've seen that kind of thing, spider goats, you kind of don't forget that."

For Cheryl, me, and many others, spider goats travel as unusual animals and technology in our imaginations. Cheryl reflected on her gratitude: "And so all of us who work on spider silk around the world probably benefited. So if I'm being funded to do basic research on spider silk, I have to think of the goats, because it's why a lot of people might have heard about it before. It might have been that news about spider goats."

As Cheryl said, the goats have been unwitting cheerleaders for golden orb weavers—and arguably all spiders—in creating an unexpected genetic splicing that has brought spiders into human consciousness in new ways. For me, personally, I have met a new animal, awed by its resplendent body and its stunning webs. But for human innovation, spiders (and humans) have hijacked goats' capacities to express and proliferate spider DNA in the form of silk protein in their milk. As the spiders have been silked, so will the goats be milked.

2

The Gifts of Goats

Milking Them for All They Are Worth

Goats just don't do well in confinement.
—Justin Jones

When I was thirty-one, I gave birth to my first daughter, Grace. Persuaded by popular and bourgeois mothering trends in San Francisco in the late 1990s, I planned to have her at home, but ultimately I transferred to a hospital for a vaginal birth with vacuum assistance. I coslept with Grace for six months and nursed on demand, even as I found breastfeeding very difficult, especially when I developed a breast infection three weeks in. Having predictable first-baby anxieties, despite our pediatrician's reassurance that some weight loss was normal in the first week or two, I felt like a failure at her initial appointment on learning she had lost weight. I doubled my efforts to feed Grace whenever she stirred, and I started to wake her more often. She is not, and never has been, a very demanding person. But I had to fatten her up.

More than twenty years later, I still laugh remembering how I shoved my nipple into her tiny mouth at any opportunity, begging her to gain weight. On growth charts, she was very tall but slender, so I continued to worry. When she was about four months old, I started my first job as an assistant professor, and despite my best attempts at pumping, I had to supplement. Grace rejected formula repeatedly, and my midwives suggested we try goat milk. I hesitated, fearful of giving her anything but my breast milk, but they insisted, explaining that goat milk had less allergens than other milks and formulas had, that goats were closer to humans than cows, and that I could combine goat milk with my breast milk easily.[1] So I agreed, shaking up foamy mixes of my milk and goat milk. Grace loved it and began to thrive and gain weight, and when I decided to wean her at ten months, the process was effortless. Goat milk came to our rescue and was life-sustaining. And even though I knew that in some divine scheme, the goat milk was not intended for Grace (it was for the goats' kids), I experienced no moral ambivalence in taking it.[2] Amalthea, a goat wet nurse of Greek mythology, fed and nurtured Zeus and became the star constellation of Capra.[3] If it was good enough for the supreme god of ancient Greek mythology...

I have often thought about how my breast milk, and human breast milk in general, is not typically used for other nonhuman animals. In contemporary postindustrial contexts, most humans don't discuss or advocate for donating our breast milk to goats or cows, and to do so would feel as if it were breaking some social taboo.[4] Historical evidence suggests that humans did nurse other species, with legal arguments proposed to change contemporary rights to interspecies breastfeeding.[5] The physician Samuel Radbill has extensively examined the practices of human nursing of animals, including goats, dogs, monkeys, and bears.[6] Geographers Frederick Simoons and James Baldwin argue that the human breastfeeding of other species was nearly a worldwide practice and that the practice diminished as Western values proliferated and dominated.[7] It seems as humans move further and further away from agrarian lives and the everyday enmeshment in interspecies relationships, the more objectionable the human breastfeeding of animals becomes. The idea of breastfeeding a baby goat brings revulsion or signals some kind of perverse kink.

Paradoxically, from my vantage point, nothing could have been more wholesome than feeding my infant daughter, her pursed rosebud lips working the bottle nipple, milky spittle dribbling down her cheek, all with the help of the happy, springy goat I conjured in my head. That mischievous rascal nibbling on clumps of grass in a sunny, open pasture came to mind every time I mixed up goat milk bottles (with the percentage of my contribution decreasing each time). But now I wonder, how did human beings get each individual goat's milk? Under what conditions do humans and goats labor? Having witnessed hours of goat milkings, done for spider silk protein milk, my nostalgic visions of goat-human collaboration for Grace's nourishment have become less bucolic, less innocent, and more vexed.

I use this story as a jumping-off point to highlight my own personal relationship with goats for their milk production, sharing the images in my head and assumptions I had when I came to the process of milking goats. Goats have been extremely useful to me throughout my motherhood journey. They literally helped me keep my baby alive, and my image of them as happily helpful mammals is part of the selective speciesism that goats benefit from, garnering more human positive attention, including my own. As I approached these goats in the field, and as I write about those experiences here, I see my ideological maternal programming, my selective speciesism, and my fondness for goats.

Additionally, I must register my own relatability to goats as the recipients of material from genetic donors. These genetic donors, the spiders, seem to recede in the collective consciousness of the popular telling and retelling of spider goat milk production. I too am a recipient of donor genetic material. And like the obscured spiders, the donor recedes into the background. I feel the connection, the spider goat and I then become the system to generate new objects—kids (for me and the goats), maternal milk (for both of us), and protein (for the goats). This similarity again imbues me with a fondness and perceived connection with the goat as entangled with some maternal ideologies that lurk in my semiconsciousness, an affective stance that infiltrates the field. Beyond connecting with the goats as other motherly bodies, my sentimental entanglement with female goats (spider goats and other goats) arouses my maternal concern for their captivity.

In this chapter, I describe goats in their coevolution with humans and the ways humans have conceptualized them as friendly, playful, and kind companions, even as we have milked and eaten them, used them to clear land and calm horses, clothed ourselves in their hair and skins, crafted ornaments and instruments from their horns, and used them for pharmaceutical creation. I then turn to the spider goats themselves for the remainder of the chapter. Many humans have a deep affection for goats, as expressed in popular culture, so, not surprisingly, there are many claims about the general health and well-being of these university spider goats. Instead of simply investigating these humans' claims about goats, I use thick description, photographs, and field notes to capture aspects of the goats' experience and the ways these animals are systems in the creation of spider silk. Using an animal studies perspective, this chapter works to reposition the animals. Intraspecies mindfulness, a practice of speculation about nonhuman species, strives to complicate anthropomorphic reflections as a means of empathetic understanding. It is an attempt at getting at, and with, another species from inside the relationship with that species instead of from a top-down relationship of domination. I conclude with the realization that spider goats are in the twilight of their usefulness to research scientists, being culled and composted (described later in this chapter) because the cost, size, and variability of goat production and reproduction make them a less useful system for silk creation than was initially hoped.

Domesticating Each Other

Goats are domesticated mammals that have lived with humans for more than ten thousand years. Remarkably, goats and humans have coevolved—we have become human in particular ways in direct interaction with the ways goats have become goats. We breed them to adjust to various terrains, encouraging their behavior as nomadic grazers that adapt to various landscapes and climates. Goats enable us to breed new generations of humans, expand civilizations, cultivate land, expand culture, and establish genealogy. They are in many ways adjacent to companion species such as cats and dogs, rather than the animals we despise.[8] Goatscaping—a play on landscaping—is the practice of employing goats to clear brush and overgrowth on land. During the recent quarantine in New York City, goats, called "four legged weed whackers," were rented to safely clean up urban parks.[9] We rely on goats as portable mammals that provide us with resources, and in turn goats are transformed by our management of their breeding and our definition of their well-being. A "healthy" goat, as retrospection on my Himalayan experience conveys, is often a goat that is scrupulously bred, dipped in chemicals, and herded long distances through diminishing grazing lands (or even transported by pickup trucks across interstates to do grazing jobs).

As social animals, goats prefer company and live in herds (in the wild, herds include about twenty animals). A doe (or nanny goat) goes into heat every twenty-one days for up to two days, and some dairy goats are seasonal breeders (influenced by daylight) whose ovulation goes dormant a couple months of the year. The gestation period is about five months. During rut, the mating season, bucks (or billy goats) produce more testosterone, emit a musky odor, and start urinating on their own legs and heads. Having smelled bucks and watched them turn up their front lip (called flehman response) as they try to inhale pheromones, I immediately grasp how they came to stand in as mythological representations of male sexuality run amok in the figure of the satyr. Goatishness is, after all, synonymous with lust or lechery.[10]

Genetic data suggests that the ibex is the ancestor of goats. There are more than three hundred species of goats, with eight primary dairy goat breeds: Alpine, LaMancha, Nigerian Dwarf, Nubian, Oberhasli, Saanen, Sable, and Toggenburg.[11] Goats are herbivores and ruminants, with four stomachs fermenting and regurgitating their food, rechewing it, and digesting it with help from microbes in their guts.

The UN estimates that more than 950 million goats roam the planet. According to the USDA, the inventory of all goats and kids in the United States on January 1, 2019, totaled 2.62 million head.[12] Goat farming in the United States is growing because of increased demand for goat products. Besides being raised for their milk (used for cheese, lotion, and soap), goats have been slaughtered for their meat (called chevon or cabrito, the most widely consumed meat in the world), used for clearing lands through their celebrated appetites and browsing skills, raised for calming horses, and sheared or plucked for mohair, angora, or cashmere. Goatskin makes parchment and can produce a supple leather. There are many uses of horns as instruments, buttons, and jewelry.[13] The pharmaceutical use of goats is also significant and explored in greater detail later in this chapter. Throughout our history, humans have made their material livelihoods on goats.

There is also a moral traffic in goats, as they are part of a global charitable economy. When my daughters were little and received an allowance or other gifted money, our practice was that they could spend a third, save a third, and give a third to a charity of their choice. When Grace was around nine years old, she and I were persuaded by the narrative and photographic images of female-headed, Black and marginalized other families rescued out of poverty with the help of a goat. So having saved enough money, Grace decided she wanted to purchase a goat for a family through Heifer International. Mail addressed to Grace began to arrive from the organization (and continues to this day), with glossy, slickly designed pages of children in brightly colorful native dress herding goats. Remembering back to that time, I think of how Grace initially felt proud and then wanted the mailing to stop because these magazines started to make her feel guilty. Fast-forward thirteen years, and most people have become more aware of the troublesome practices of gifting agricultural animals.[14] Noting the lactose intolerance of some ethnic groups, the lack of sustainability in giving animals to people with limited resources for animal care, and the displacement of indigenous agriculture, many investigative journalism stories suggest that these Global North practices actually increase suffering in the Global South.[15] "White guilt fucks shit up again" might be a better title for the magazine than *World Ark*.

Unlike depictions of spiders as fearsome, dangerous, or evil in many children's picture books, normative constructions of goats represent some sort of docile, content, friendly, unassuming creature.[16] As sociologists Arnold Arluke and Clinton Sanders groundbreaking work argues, the way we regard animals is deeply entwined with our myths; our very meaning-making practices deem some animals as good, worthy, and deserving and others as dirty, evil, or vile.[17] For us human beings to treat animals differentially, as these scholars demonstrate, we must develop flexible and movable boundaries between ourselves and other animals and between species to develop the justifications for our violence.

Our early anthropomorphic training grounds attribute human characteristics to goats (playful, social, silly), easing the strangeness of our use of their milk for our own consumption. They are similar to us. But when we wish to slaughter them for meat, their differences must be accentuated so we are not skirting too close to cannibalism. One implication of this ability to flexibly shift between sameness and difference leads to a lack of consideration for the goat's own subjective experience and a lackadaisical attitude toward the goat when it is used for transgenic means.

Saanen Goats to Spider Goats

The spider silk goats are a transgenic form of Saanen goats (*Capra aegagrus hircus*) goats. Chosen because they are among the higher milk producers with the lowest percentage of milk fat, Saanen goats produce milk that is easier to filter for spider silk protein (see chapter 3). These goats were imported into the United States between 1904 and 1930 from

Europe.[18] The milk I fed my daughter Grace also came from Saanen goats, among other dairy breeds.[19]

Interested in speaking to someone outside the Spider Silk Lab about Saanen goats, I contacted John White, the registrar of the American Dairy Goat Association (ADGA), the agency that maintains the breed standards for dairy goats. For several years, John served as a linear appraiser for ADGA (a person who evaluates sires and dams and assesses their quality) while also working on goat dairy farms.[20] He said that Saanen goats were one of the most popular breeds in the United States because "they are stable and solid and give a lot of milk and don't get excited about much. They work well in a big setting." As he was curious, we discussed my project and my description of the spider goats as friendly. Quickly he agreed. "Yes, I could see that these goats can handle a lot of lab testing and blood work. They aren't going to be too flighty." Interesting word choice, I thought, as *flighty* is a pejorative strongly associated with the feminine, bringing to mind a woman who indulges her sudden whims and frivolous desires. I wonder if male animals are ever called flighty.

John then explained how the ADGA registered animals, saying that registered animals generated more value for the human owners.

"Why would someone want a purebred goat?" I asked.

"With respect to milk production," he said, "you just don't know what you are going to get with random mutts or crossbreeds. You don't know if they are going to be good milkers. A breed is mostly predictable, and that's useful to know."

I asked about the goats' life expectancy.

"Saanens—I'd say about ten years, where seven or eight are productive years, depending on how much you freshen them."[21]

As John explained the relationship between breeds and milk production, I thought back to my fieldwork in Utah and the goats I had met there. A majority of my observations focused on milking the goats.

I remember pulling into South Farm and driving past the lone portable toilet and a small stable of handsome black cows—reportedly a herd for heart research. The last barn, where the goats live, is a metal and cement structure about twenty-five by ten yards, divided into different pens on the sides, with a long, wide aisle down the middle. A manger full of hay is at the front of each pen. You can hear the goats before you see them. Loud bleating almost like babies crying and whining. The goaty smell wafts into your nose as you enter their barn. Musky, dank, thick. Each pen holds different goats, most of them pure white with white wispy beards and yellow-brown eyes with bizarre rectangular pupils. Many wear collars or chains around their necks. Two bucks stand close to one another, segregated from the rest; they move toward the fence as I enter, gently rubbing their heads against each other's necks.

The does live in other pens, and each animal has an ear tag; some goats are spray-painted with neon blue marks to indicate they have been RFID chipped.[22] Transgenic goats milked for spider silk protein are kept separate from nontransgenic goats, milked for feeding kids. As a greeting, I grab hay to feed them by hand as they approach me at the edge of their fencing. I like the feeling of their noses exploring my cupped hands and the way they climb up on the edges of the fencing, stretching their necks. The goats have strong lips that purse out and pull back. They nibble at my fingers but don't bite me. My fingertips get wet with their saliva. Through this long barn is a separate structure called the milking parlor. A temporary pen stands in the middle of the parlor, where goats wait their turn to jump up onto the milking platform. The goats seem completely obedient to this twice-a-day routine and almost seem eager to be milked. Is it their full udders or the molasses-covered feed that they get only during milking? Or is it just the breakup of the boredom of being in a pen all day long except for these twice-daily twenty-yard excursions? Perhaps it isn't so much obedience as the realization these goats are technology that is built to be unable to resist?[23]

Two stands are constructed for milking the goats. A metal lever immobilizes their heads while suction cups are placed on their teats. Feed is poured into containers in front of the goat's faces, and the suction compressor is turned on. Before the suction cups are connected, the herdsman wipes down their teats with an antibacterial wipe. The noise of the machines is loud and grinding, but the goats just eat their feed, seemingly unfazed. I can't help but be reminded of hours of pumping my own milk in faculty offices, counting the ceiling tiles above my desk while my undergraduate students waited outside. After the milk stops flowing, about ten minutes, the goats are disconnected and their teats are sprayed with Fight Bac (a disinfectant to prevent mastitis in dairy animals). I sympathize with this preventative measure, as my breasts

Figure 2.1. Different pens of nanny goats at South Farm, some transgenic goats and some nontransgenic. Photo by Lisa Jean Moore.

needed special care too. I recall applying lanolin, a wax from sheep's wool, to my sore and cracked nipples after nursing and pumping to comfort and protect them. The milked goats are then either walked back to their pens or allowed to find their way to the pens themselves. The few stragglers are rounded up at the end of the milking. At the back of the milking parlor, the herdsmen hand-wash the equipment at a utility sink and dry it on the racks of a broken dishwasher. A couple of refrigerators store milk before transport to the lab's freezer rooms. Two barn cats wander around looking for spilled milk, which they lap up. At the edges are two small pens for the youngest kids, curious about all the happenings in the parlor. The kids are undeniably cute as they bleat, jump around, and shake their butts.

For my whole life, I've been enrolled to think and feel about mammals a certain way, and I have passed these attitudes on to my children. My emotional connection to reproduction, a feeling I also transmit to my daughters, comes from historically sex-binary thinking. My own practice is frustrating to me, as I have criticized this sex binary, but I have been effectively socialized by it and socialize my children to it.

Figure 2.2. Goats waiting to be milked. Photo by C. Ray Borck.

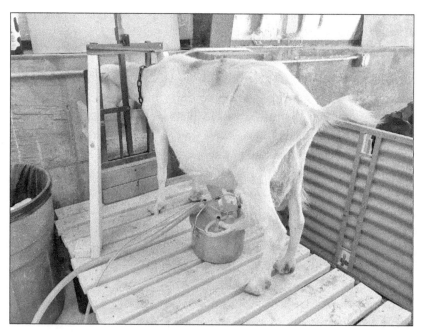

Figure 2.3. Goat number 602 getting milked (she was born in 2015 and was the second goat birthed that year). Photo by C. Ray Borck.

Western white womanhood and our tendency toward sentimentalizing has reaffirmed the sex binary, argues Kyla Schuller.[24] My ability to be impressed by representations of maternal intimacies and the capacity to feel sympathy and to know what I am feeling is instilled in me through a historical legacy that encourages and fosters white women's feeling and expression of feeling. Schuller, through a skillfully analysis of nineteenth-century texts, reveals the myriad ways the maintenance of binary sex differences was a racist project of white supremacy. She uses the term *impressibility* (as distinct from *impressionability*), defined as "the capacity of a substance to receive impressions from external objects that thereby change its characteristics."[25]

My very self, a legacy of white womanhood, constrains me to be emotionally malleable—I am and choose to be impressed by mammals who mother. Stories of how to become "a mommy" disciplined me and how I interacted with the world, and—beyond me—I impose it on my children. I model for them sympathy for mommy animals; this modelling is a biopolitical and technical act of sentiment that I was taught and that I teach. Reading stories of mommy otters swimming with babies on their bellies basking in the sunshine, watching videos of mommy whales mourning the loss of their babies by pushing the babies' bodies for miles in grief, observing mommy monkeys holding their babies close to their nipples for easy feeding—these are the discursive representations I was brought up on and shared with my children. I have been encouraged to know some universal mammalian animal affects of maternalism, and it is not by accident that I've internalized maternal tendencies because I am "properly" gendered and raced. I have expectations of the goats as part of this discursive training I received and delivered.

As much as I attempt to free my mind of essentializing universal claims about mothers and mothering, I catch myself slipping into what mothers should naturally do. But here, at this farm, all these moms walk past these babies. The babies see the moms, but it's as if they don't really know each other. They could be a different species. The babies are screaming for food, but they're not screaming at the moms. They're screaming at the humans because the kids think the humans have the food. And actually, the humans do have the food. But they took it from the goats. Thinking back on this as I spoke with John White, I hesitated and stumbled over my words. "I know this is perhaps going to sound crazy to you, and I'm not even sure how to put it. But when I was in Utah, I was sad to see these young kids that had been taken away from their moms and I wondered if it was sad for the goats and . . ." I trailed off unsure of how he would react to my question. John replied, generously, "I understand you." He continued, describing the goats' reactions from his experience working on dairy farms:

> The goats who have never gotten to be moms are fine. They haven't been a mom. Neither has their mom, r her mom all the way back. They actually don't know what to do with the kids. They were basically raised for generations for it to be normal for the kids to be taken away, and they don't react at all. They just go back to the feed. However, if they do get a chance to be a mother, then it is more difficult for them. Once they get the experience of being a mom for more than two or three days, then it can be agony for them to have their kids removed. They just stand around and cry and fuss. You know they have that hormonal aspect of having had a kid, and stand at the fence and scream all day.

This sounded awful to me, and I was impressed that John admitted to me that the goats could feel sorrow. I told him it made me feel sad. He replied, "Most dairy farms kids are taken away at birth—because it causes more ruckus if you don't, and when you are freshening several hundreds of them, you can just put them back into the system, and nothing more is ever said about it. They don't think a thing about it." And while I hear him and his expertise, my beliefs about what a mommy does run amok.

Is it wrong to make these assumptions about goat subjectivity? We can speculate about when they do care, when they don't care, what they like or dislike, whether they mourn or forget. Clearly, spiders are mothers too, but we never consider their maternal instincts. Maternal bodies—humans, goats, spiders—must always remain flexible to scientific intervention, infant theft, and lactation surveillance.

While I was conducting this research and writing this book, there has been a pull to consider the bioethical implications of the scientists' actions. To be completely up front, I don't feel particularly judgmental of the scientists for their treatment of the goats. As I have described and will continue to describe, I don't think that these goats' lives are ideal. Given the continuum of all goats' lives on the planet, and considering

Figure 2.4. Kids about five months old (not yet tested for transgenes) waiting for a bucket of nontransgenic milk. Photo by C. Ray Borck.

my very humancentric evaluations of what is a good goat life, I don't think these goats have the worst lives, either. Of course, if I were a goat, I would probably prefer to be free, roaming on the island of Capri and climbing the cliffs. But having never been a goat, I am not entirely sure what sort of life I would be signing up for with that fantasy.

As I have discussed and debated the work with colleagues and friends, I have felt defensive of the scientists and the herdsmen because they often get cast as some evildoers. One conversation with my boyfriend, who joined me on parts of my fieldwork, pushed me toward an epiphany.

"What is insane about it isn't the way they are treated," he said. "It is insane to me that they don't know they are suffering. They are like us."

I asked what he meant.

"You know, like, they have false consciousness. And it's depressing. They don't know there is a better life, because they have never been outside that enclosure. They don't know how to miss running free, because they have never run free."

My previous work has engaged with this contemporary Western tendency toward anthropocentrism in our thinking about and acting toward the more-than-human world. Anthropocentrism is a belief system or way of thinking that regards humans as the center of all existence, above all other living things. We use our very human experiences to selectively narrate things for animals when it suits us. This perspective can also slip into thinking of nonhuman beings as inferior to humans, and as a result, anthropocentrism is a form of speciesism akin to racism, sexism, and classism. In other words, anthropocentric thinking requires the same stratified value judgments as when we see certain types of people (men, white people, able-bodied people) as more worthy than other people (women, people of color, or people with disabilities). Some critical animal studies implore us to challenge our anthropocentric tendencies through our scholarship and other practices. But at the same time, our own positionality as sentient humans can be mined as a source of attachment and engagement with other species.[26]

In this vein, anthropologist Barbara King insists that we value animals' emotions and, in particular, bear witness to their experience of loss.[27] Her work explores the dimensions of animal bereavement, analyzing how animals ritualize their dead and cross-species grief, including a case of a mother-daughter goat duo that were reunited after a separation and whose joy was undeniable. Witnessing how the kid goats at South Farm seem to have little regard for their mothers makes me wonder what humans have done to these goats. To push King's observations a bit further, what if grief isn't possible, because you've never experienced connection? Have we robbed kids and nannies of their capacity to grieve? What is lost when a species can never experience grief since the affective bonds needed to grieve are denied?

Throughout my research, I saw clear examples of humans' tendency to toggle between empathy toward animals, specifically goats, and dismissal of them. Also, goats are habituated to human-based priorities that make goats for-human prerogatives rather than for-goat prerogatives. Are goats so relaxed in their pens because they don't know that a certain kid is their baby? They return to their pens after milking because that is what they always do. They are produced to not know certain things, and other options of life are not accessible to them as a means of keeping them docile and predictable and obedient. They comply with their lives because it is the only life they know, the only option available to them, the only choice. For my boyfriend, this is a sad reality and a statement on human suffering as well. I shudder realizing how right he is. Perhaps decreasing animal *suffering* is the bare minimum of our human ethical responsibility. We manipulate the conditions so that their suffering never has the opportunity to emerge. The suffering thus may have been *technically* removed, however the material conditions that would normally produce goat suffering persist. In turn, during the routine of their everyday life as living machines, their human captors, keepers, or administrators may appear less cruel, even ordinary and benign.

Where Do Spider Goats Come From?

Transgenic animals were first developed in 1985, used for biopharmaceutical innovation to synthesize drugs. But the mammals' ability to create certain desired proteins in their milk in sufficient quantity and quality was quickly exploited. Milk is also a renewable resource, easily taken from animals, with limited harm to them. The ideal transgenic animal produces plenty of protein-rich milk and has a relatively short generation time; in other words, a female goat takes roughly 150 days to kid and can then be milked for up to ten months. Typically, dairy animals are used because of their high volume of milk production. Goats have been selected for several reasons, the chief of which is the short generation time of eighteen months. One animal produces about eight hundred liters of milk per year. Using goats has been shown to dramatically increase the yield of the active protein by more than ten times that produced from cell culture.

As discussed in journalist Emily Anthes's book *Frankenstein's Cat*, genetically engineered goats were created in 2006 to produce an anticoagulant for individuals prone to clots in their legs and lungs.[28] This process is similar to the transgenic science involved in producing spider silk protein. After scientists injected a gene for human antithrombin, the anticoagulant, into fertilized goat eggs, the eggs were implanted in a goat's womb. A founder female, one that breeds kids, is created. Of the kids born, some were transgenic. The antithrombin gene was paired with a promoter (a DNA sequence that controls gene activity) in the mammary glands. When transgenic goats lactate, this promoter turns on the transgene, and goats udders fill with milk that has antithrombin. The goats are then milked, and the milk is purified to extract the protein.

Randy Lewis, the director of the Spider Silk Lab, sent me this email in response to my query about the origin of his herd of spider goats. I quote it at length to capture the complexity:

> The full story goes as follows. The CEO of Nexia Biotechnologies in Canada had connections to the Animal Science Dept. at the University of Wyoming and was visiting them. Our department was in the same building and the head of the AS Dept. suggested we meet and we did. After a substantial conversation, we continued to explore the use of their goat system to produce proteins. They had already successfully done one protein and were looking to expand their products. So it was relatively easy for us to send them the cloned DNA we had for each of the spider dragline silk proteins to use in their transgenic goat system. In a little less than two years, they had produced the first "spider" goats. Before that they had made transgenic mouse mammary cells and demonstrated that they produced the spider silk proteins and secreted them which is what would be needed for the goats to work correctly.
>
> Goats were used by Nexia as their gestation and time to first ovulation after birth were much shorter than other milking animals. Thus cutting the time to production by over half. In addition, for reasons still not clear, cloning goats (which is essentially the same as creating and implanting the transgenic embryos) is highly successful compared to cattle. Goats also produce about the same amount of milk per pound of feed as dairy cattle and are much easier to handle. And finally we laughingly tell everyone that baby goats are much cuter than baby cattle.

After the scientific mechanism was understood, a series of deliberations preceded the choice of goats as the right tool for the job.[29] These considerations included the speed of goats' reproductive capacities, their dispositions for ease of human handling, a touch of mystery over the success rate of cloning, and their emotional appeal to humans. Goats were about to become a living factory for the production of spider silk protein.[30]

In the 1990s, the US Department of Defense offered financial support to a now-defunct Canadian biotech company, Nexia Biotechnologies, as the company engineered a herd of transgenic goats. The goats were part of a project attempting to produce new fabrics both strong and flexible—for wearable armor, bulletproof vests, or pants that repel shrapnel. When the project collapsed in Canada, Randy acquired the herd and eventually brought them to the western United States. Here he continued to do research with the goats; the work was funded by a variety of military and other federal agencies. On my preliminary research trip to his lab, he explained, "At first we got money for basic research from the Office of Naval Research. Then we ended up at the army, and the army had it for a while, and then it went to air force. And now we have made a complete 360 and we are back at naval but more for our adhesive properties rather than mechanical properties. We have also had funding from NIH, NSF, DOE [National Institutes of Health, National Science Foundation, US Department of Energy]." Clearly, the various funding streams flow into one another.

The circuitous but also dependable funding creates the financial backing for research on spider silk. One branch of the US military has continuously stepped up to fund the lab work on spider silk. Even if the goats are not currently the most profitable (or affordable) way of generating silk, the goats are the celebrities of the lab. One scientist described them as "really good for attention and publicity because people love goats," implying that successful grant funding is assisted by charismatic animals. The story of the goat transgenic manipulation is the most compelling to come out of the lab, possibly because the goats are vertebrates and mammals.

While the US military has been the primary driver of transgenic spider silk research, and although these goats exist because of military funding, the military's investments and interests diverge from those of the scientists.[31] The military wants to find lightweight fabric technologies to protect its armed forces from weapons and wartime risks. The scientists want to create such body-compatible medical insertable products as prosthetics and bandages. While neither potential of spider silk materials has yet to be realized, the laboratory is in full pursuit of both—balancing the economic value of the goats for military actors with biomedical innovations for health care.

The ownership of the herd is more tricky to sort out than the story of how they came to be. I asked Randy to explain. "The exact ownership of the goats is not entirely clear," he said. "The agreement with the company that made them has the agreement with the University of Wyoming, who has an agreement with Utah State University." This question of the goats' ownership has moved with Randy from one university to another. The trademarking and ownership of the gene construct more clearer, he said: "The gene sequences are patented through UW, which has licensed them to USU" (see chapter 3).

Spider Goats as Protein Systems

Coincidentally, while I was conducting this research, Perumal Murugan's Tamil novel *The Story of a Goat* was published in English. In 2014, Murugan's previous work was attacked by conservative religious leaders and politicians in India because he depicted extramarital sex in his book *One Part Woman*. Protests demanding a banning of the book and featuring book burning ensued, and Murugan was forced to sign an apology. *The Story of a Goat* is his first book since the controversy and his literary silence. In the preface, the author writes, "I am fearful of writing about humans; even more fearful of writing about gods. . . . All right, then, let me write about animals."[32]

The book, a parable of Indian village life, is narrated entirely through a goat's devoted relationship with an old woman. It is also a political allegory about violence and suffering through oppressive regimes. The *New Yorker* review by Amitava Kumar describes the book as the story of "a female laborer caught in a reproductive economy in which her experience of love is real yet fleeting, her voice never silent and yet unheard."[33] Murugan writes about the ordinary life of Poonachi, an all-black goat with a prophecy foretold by a mystical human. Throughout the novel, Poonachi is yoked, raped, and bred, as she bears witness to gruesome killing. She is a self-aware animal, and her reflections, seemingly simple, are commentaries on social power. A conversation between her and her beloved, Poovan, about mechanisms of human exploitation of male and female goats, Poovan explains his circumstances:

> "Death can come to a buck kid at any time. We die for meat. We die for sacrifice. I live for moments like these, when I get to be with you, even if only by chance," Poovan said.

Poonachi replied, "Do you think a female has it any better? It's better to die than to go through the ordeal of birthing and bringing up kids."

Here the goat's life is at once the story of controlled cultivation of animal bodies and a metaphor for human experience, helping me think through some of the gendered ways mammals are exploited (and desired) for their reproductive capacities.

To produce the protein, the goats must be bred to make the next generation of goats as well as milk. When the nannies give birth, researchers check the kids for the presence of the transgene by using a polymerase chain reaction technique shortly after they are born.[34] The transgenic bucks are culled. The scientists have found when they mate a transgenic nanny goat to a transgenic billy goat, the female offspring that are transgenic will sometimes produce too much protein in their milk. The high protein concentration can cause some aggregation in the udder and result in mastitis, or infection of the mammary glands.[35] So transgenic does are bred with nontransgenic bucks in a process called heterozygous reproduction. Since the lab only needs two males to reproduce with the herd, the lab can sell the nontransgenic bucks as kids to ranchers outside the university system (as these animals are like any other goat and have no transgenes in their bodies). South Farm's young male goats are not used for breeding, to avoid inbreeding of goats. The nontransgenic males that are sold are called *wethers*, or castrated males. With wethers, there is no way for any possible recessed or unexpressed genes to transmit.

Justin Jones, director and a principal investigator at the Spider Silk Lab, told me about the current bucks' lineage. "They were purchased, and their lineage can be traced through Drake Farms in Utah and California. They're a very large dairy goat farm, and they throw very good does and very good billy goats, and you've got years and years of lactation records behind those billy goats. So you kind of know, the kids that they throw are going to be very, very good milk producers."

This phrase, *throwing males or females*, was at first jarring to me. I imagined a human squatting down in a catcher's ready stance with a huge fluffy mitt, prepared for the female's vagina to throw out a kid. But the term *throw* as applied to animals has a peculiar history. Throwing refers to animals from the 1820s and is related to *throw back*, meaning "to revert to an ancestral type or characteristic." So *throwing* in this context could be connected to tracing lineage or ancestry.[36] As I ponder this, I consider how I've used *throwback* to talk about a style of clothing, a haircut, or a cultural object. I use the term in connection to time and repetition of the past brought into the present. I've never used it to discuss the passing on of genes. As in when I threw my daughters Greta, Georgia, or Grace, they were throwbacks to my Granny Jean. Instead, I've been taught that I birthed, or rather the doctor delivered, babies through my body. For humans, this notion of delivering babies is as if the delivery happened to the woman, a package received or even as if the baby is delivered through the woman to someone else. I might rather imagine that I have thrown my girls.

The attributes of a good buck are being a good stud and reproducing females, transgenic goats, and good milkers. So to breed good kids is to throw consistently good milkers based on the traits of their mothers and their mothers' mothers. Billies are good, Justin says, until their "teeth start to fall out, and they get too skinny." I asked Justin if he was happy with these bucks, thinking back to how seemingly affectionate they were with one another and how handsome they were.

"Yes," he replied, "those two males have done well. We've had in excess of 90 percent conception rates; we're throwing twins and triplets. I think we averaged in our last two kidding cycles—we average twins, which is a bit unusual for us, because we do tend to have younger does coming along that drop that number because they tend to only throw singles and they've also thrown a fairly high percentage of female offspring. So they've both done well. Have they done extraordinarily well? That's hard to say. Have they done better than billies I've had in the past? Absolutely, yeah."

Justin's measurement of a stud's worth is a fairly simple quantitative calculation of how many female kids the billy can make.

In the fall of 2019, the herd count, according to herdsman Amber Thorton, stood at twenty transgenic and twelve nontransgenic goats. She further broke down the numbers: "Of our transgenic goats, twelve are MaSp 2, five are MaSp 1, and three are not yet typed." The herd was made up of mostly mature does and yearlings and included three spring kids. "The nontransgenic goats include ten mature does or yearlings and two mature bucks," she said. As previously noted, transgenic goats produce one of two proteins from the dragline spider silk, either MaSp 1 or MaSp

2. Historically, the goats had produced about one gram of protein per liter of goat milk. Justin said that this group was the fourteenth or fifteenth generation of goats.

As Randy Lewis was nearing retirement, Justin was preparing to take over leadership of the lab. A warm man who brings to mind Western rugged sturdiness, Justin speaks with a straightforward familiarity, a slight drawl accenting his speech. Observing him around the lab, I can plainly see that the undergraduate students, graduate students, and postdoctoral fellows all admire and respect Justin. He interacts with them easily, sharing many inside jokes and self-deprecating asides.

Justin explains the variability of goat protein production: "We had goats that were doing two to four grams per liter. So we scaled down the number of goats we have in that herd with the premise that, you know, we don't need as many goats, right? And then our star goat got sick and died on us. And so the other ones were still good. They're just not that star goat where we can purify—you know, we were putting ten, fifteen, twenty grams on the bench each time we ran a purification. And now we're putting, you know, two to four."

Who is blamed? I wonder. I ask, "Is that the goat's fault?"

Justin is emphatic. "No, no, no, no, that's not their fault at all. This is just part of the problems working with a biological system. Combined with working under the tight constraints we have. We can have bad luck with the goats and bad luck with the purification."

The spider goat protein production is an assemblage of goats, specifically their anatomy, their feed, their genealogy, the environment, humans, the goats' physical labor, animal husbandry, technical ingenuity, engineering, hardware, and the purification system of tubing, compressors, chemicals, vials, and filtration. With so many working parts, there are several locations of potential error. The spider goats are approaching three decades of producing spider silk protein for scientists. But the scientists have needed to learn about the vicissitudes of goat milk production and reproduction. The researchers want to get the most silk protein from the milk as possible and to reproduce transgenic kids from pregnant nannies.

But these goals have required fine-tuning processes. Justin recalls increasing the transgenic female numbers in the herd: "We've got better over the years at maybe manipulating, not necessarily the percentage transgenic but the percentage that are female and transgenic, which is the percentage we really care about. At the University of Wyoming, our odds were very low. Mendelian genetics tells us we should end up with 25 percent females that are transgenic. And we were running, you know, 15 to 20 percent because we weren't throwing very many females. For whatever reason, we would throw a lot of males."

He concludes, telling me, "And now we throw more right on that line of 25 percent if not higher." He sits back and grins.

"What have you done to make it higher?" I ask.

"It's an interesting story, because Randy and I visited a farm in South East Wyoming, and there was a lady there. When we told her our problem of throwing predominantly male, she said, 'Oh, there's a trick that you can solve that, and that's to feed vinegar to the females. Prior to, you know, insemination.' And we were like, 'Whatever. That's not, that can't really be true.' But anyway, we got, you know, our odds continued to be fairly low on the Mendelian genetic side of things. So we started feeding vinegar, and the number of female kids that we threw went up."

He looked at me as I raised my eyebrows and snorted. "Really?" I said. This old folk remedy delivered offhandedly by a woman in Wyoming becomes part of the folk knowledge of animal care not often documented in scientific journals?

Justin says, "It's supposed to acidify the vaginal tract and thereby select for females. Yes, it's her environment. It seems like it's working. But there's also, you know, other factors. Of course, during that time, you'd be bringing new billy goats and other things. But with that said, I've never stopped feeding the vinegar, OK? And it's kind of funny, the goats just absolutely love it. It's like salad dressing for their feed."

I laugh and think of all the ways I've tried to disguise distasteful foods for my toddlers. Tricking them into eating something for some nutritional outcome under the purview of the good mother. Justin is thrilled to have this emotional connection where the goats seem to enjoy excitement. "You know, you always try and provide enrichment for your animals, and that's one of the most enriching things we do. They get excited about seeing you walking out to the barn with a jug of vinegar. And I get up and I stand on the fence, and they bleat at you, and you can smell the vinegar and alfalfa on their breath and they just love it."

Continuing our conversation, I ask Justin about how the goats were managed to produce spider silk in the optimal amount.

Justin explains, "Throughout the lactation cycle, the protein production is going to vary to some extent—how much, I don't know." He describes the various contingencies that will relate to goat milk and protein production: "When the days get hot, protein production tends to drop a little bit. You know, we've had instances where goats get stressed like you have a tour go through the farm and somebody brings their dog. Goats really don't like dogs, it really stresses them out, you'll see milk production drop and protein production drop."

Identifying why goat milk protein production drops is not an exact science. Because the Spider Silk Lab relies on having access to the protein to conduct experiments, the researchers notice the decrease in protein production. And the decrease is stressful and seems to change the way the goats are regarded. During one meeting, Thomas Harris, a postdoctoral fellow describing some experiment results, quipped, "So one dope from *E. coli* formed a spongelike substance, so strike another one against the goats."[37] Everyone in the lab laughed. To clarify, Thomas was explaining the favorable outcome of an experiment using spider silk protein from a system other than the goats. The bacterium *E. coli* had produced this positive result at a time when the goats seemed on thin ice. The mammals' reliability as protein producers was at that moment precarious.

The Spider Goat's Life (as Told by Humans)

One summer afternoon, I met Justin in his office for a scheduled interview to get a bit of his biography. Justin originally wanted to be a veterinarian, he said. "So I started volunteering with veterinarians and in Wyoming, and looking at what they dealt with on a daily basis, [I] decided that was really not what I wanted, because there was too much tragedy. You know, horses kept in a ten-by-ten pen, their intestines full of sand. It was too hard." Clearly, Justin has been driven by some types of empathy for animals. He sat back and looked into my eyes and paused. "And so I found I actually started working in Randy's laboratory at the University of Wyoming back in 1995 as a fresh-faced undergraduate. And I just fell in love with research and, more specifically, the spider silk aspect of things because I was never drawn to research for research's sake. I wanted to have a purpose."

After Justin earned his master's degree, he spent a few years working for a biotech start-up, but "that environment that they were working was using practices that weren't always conducive to doing the best work because of greed."[38] So he returned to the university and received a doctorate in biology and continued working in Randy's lab. A professor on the tenure track, Justin has been working with Randy for around twenty-five years now and is well prepared to take over the Spider Silk Lab.

I asked Justin how he felt about the everyday life of the goat. He sat back at his desk and said, "The life is pretty good, right? They are under twenty-four-hour video surveillance so that we can keep an eye on them if something happened. We can go back to the loops and look at that. I employ two people that are out there every day checking on them, doing all that animal-husbandry stuff, you know. They get access to high-quality alfalfa and feed, so the life is pretty good."

I nodded and said that it was probably better than a slaughterhouse or an industrial-sized commercial farm. At this point in the research, I was still deciding how I felt about the goats' lives relative to all other ways goats could be raised, and I didn't want to be outwardly judgmental. But I wondered how to measure their life as "pretty good" compared with other domesticated animals.

He tilted his head and replied, "They are necessarily constrained compared to other goats, because of the Department of Agriculture. To a certain extent, the USDA, the FDA, you know, requiring that we're behind double fences and we've got good control. You throw into that mixture our own internal Animal Care and Use Committee and the restrictions they place on the animals and what we can do with them. Yeah, they're highly looked after."

Justin seemed to be indicating that he would favor giving the goats more free range but that because of restrictions on transgenic animals, the spider goats were limited in their ability to roam. He then smiled and said, "And goats are wonderful, right? One reason we use them is that I don't know anybody that's ever been killed by a goat, right? They can step on you. It'll hurt maybe for a minute, whereas a cow steps on you and you have a broken foot."

But Justin is also aware that the life of the goats is not ideal. "Our animals are kept in confinement, and there's a lot of problems with raising animals in confined areas. You know, you get pink eye in one, and it's through your herd that quick. And so, more serious ailments also spread that quick." The goats cannot mix with other goats and are kept segregated from other goats at South Farm.

Justin is using emotional management strategies similar to those of other scientists working with animals in labs.[39] Patricia Morris's sociological investigation into veterinary practices provides insight into how humans manage animal death and develop mechanisms to make decisions about euthanasia. Morris observed veterinarians and found that because animals are deemed property to human owners, decisions about euthanasia with clients are different from decisions about human-based medical delivery.[40] Since animals are owned (i.e., they are property), human handlers (vets, ranchers, research scientists) cannot merely see animals as patients with medical problems. The animals also have to be understood as objects that add or subtract value from human life. For example, if the cat's surgery becomes prohibitively expensive, a human can elect to euthanize the pet, and the vet will often comply. The animal can slip from subject to object in these decisions of culling.

During my research, I spent a couple of days with Andrew Jones and Amber Thorton, the herdsmen for the goats. They share the morning and afternoon milking shifts for the herd and the tasks of feeding, animal husbandry, and general care. Andrew and Amber work with both the transgenic goats and the nontransgenic ones, as well as the kids and two bucks. The herdsmen communicate frequently with the lab to assess which goats are to be milked and for which proteins, as well as managing the cull list and deciding which goats need to be "dried out."

Amber, an undergraduate in the agricultural school, had worked with and shown goats when she was a high school student. I shadowed her on my first visit to see the goats milked, and she moved through the routines with quickness, certainty, and care. I asked her how she felt about working with the goats. She replied, "These goats are friendly because they get messed with a lot. In terms of livestock in general goats, they have a little bit more personality. So they're pretty fun to work with. And they've all got their own little personalities."

I pointed at two goats that were standing separate from the rest in the pen. "What about these two? What are their personalities?"

Amber hesitated a bit at this question. The work of sociologist Nik Taylor is instructive here.[41] Taylor's ethnographic work at animal shelters demonstrated how workers immediately name animals arriving at the shelter as a mechanism to create affective connection to the animals. Naming is one strategy used to attribute personhood and personality to animals and hence status. This naming suggests that when animals are seen as a herd or a species, it is easier to make them generic. When the animals are considered generically, humans can exploit them in the aggregate rather than on an individual level, possibly making it easier to distance themselves from moral questions about harm. At South Farm, while an occasional goat gets a name, individual goats are more often referred to as a number.

"She's kind of chill," Amber said, pointing to a goat that to my naive eyes looked identical to the others. "And this one right here, 602, you'll see when we open the gate, she knows how to get on the milk stand once we get into the parlor. She just jumps over the fence and gets on the stand every single time. So she'll do that when we go in." And sure enough, when we took this group in to be milked, this goat cut the line and just jumped onto the stand. I tried to help Amber, but mostly I just walked around and took pictures and nuzzled the goats' noses when they seemed interested. The kids stood in their smaller pens and cried at us humans, eagerly watching the bucket filling with nontransgenic milk. It would be split between two groups of kids. Once it was poured, five little heads immediately dived into the pail, with sounds of them lapping up the milk. Their little tails wagged rapidly for the thirty seconds it took them to finish drinking.

After milking the goats with the pump, Amber then hand-milked the teats to get any remaining milk out. She offered to let me try. Milking a goat had secretly been a fantasy of mine for a long time so I tried to play it cool, but internally I was thrilled. Amber put my hands about two inches up and around the warm teats, instructing me to use my thumb and pointer finger to gently tighten my grip on the teat and then gently squeeze with my palm as I squeezed, aiming down. A dribble and then a spurt of milk came out, and while it was difficult to believe I was not hurting the goat, I enjoyed the experience. It put me in mind of my *Little House on the Prairie* fantasies and I felt purposeful and earthy. The intimacy invigorated me, the experience of pulling on their nipples, the smell left on my hands.

Later, Amber also let me walk goats back from the milking area to their pen in the barn but the task really turned into my just following the goats on their brisk walk. I found myself wishing for one goat to break free, starting a madcap adventure of hijinks, but each goat dutifully took the well-trodden journey back to the pen.

Amber giggled watching me slowly get the rhythm of my milking. I told her how I always really liked goats but had never been this intimate with them.

She nodded. "I mean, I like them," she said. "I generally try not to get too attached to them. Right, but I like them." Amber's decision to maintain emotional distance from the goats is related to her being part of a team that determines the cull list I discuss below, a clear limit on the fantasy of the good goat life.

Andrew, the other herder and Justin's son, was also an undergrad at USU. Like Amber, he was comfortable around the goats and exuded confidence managing the herd. He had a slightly different style of coordinating goat milking. After milking, he let the goats find their own way back to the barn. I watched as, amazingly, they did return, and I thought of the Stockholm syndrome, where hostages develop a psychological alliance with their captors.

While he milked the goats, I asked Andrew whether he had ever received criticism or a raised eyebrow for his labor with the goats. He had a straightforward way of communicating, like his dad.

"Well," he said, "so one of the things is, they're just regular goats. A lot of people have this notion of transgenic, you know, genetically modified organisms, genetically engineered organisms, that they're abominations against God, especially here in Utah, particularly with a religious state." He moved to transfer the nontransgenic goat milk to the kids, pushing their heads away so they wouldn't get wet as he poured it into a bucket. "And the thing is, they're just regular goats. If you know anything about biology, all it is, is a certain section of DNA that translates to mRNA, which just produces a biocompatible protein, not like they're producing some sort of toxin. It's not unhealthy or detrimental to the goats. It's just a protein that they produce. I mean, spider silk, it's biocompatible. It's not even both proteins [MaSP1 and MaSP2] synthesizing in the same goats."

Both Amber and Andrew believe that these goats are no different from other goats they have encountered, except perhaps that the transgenic goats' lives are more circumscribed by regulations that confine the animals. During our conversations, Amber and Andrew both switched between a more familiar, comfortable flow of chatting about the goats to a more performative and controlled speech. These young adults (in their early twenties) were practicing being more mature adults. As students, despite their high level of intimate contact with the goats, they are clearly being socialized to be detached and objective scientists. At the same time, however, as sociologist Rhoda Wilkie observes in humans who work with livestock, people's relationship with domesticated livestock is not one of callous disengagement or one in which the animals are "undifferentiated commodities."[42] Amber and Andrew, like Wilkie's livestock workers, constantly vacillated between cognitive detachment and emotional engagement with the goats on a daily basis. They constructed ways of being with the goats, and after my initial delight in touching the goats, I started to adopt these ways as well. We can simultaneously care about, and care for, the goats while trying to get them to stick to the task at hand with the greatest efficiency.

Regulating Spider Goats

As research animals, the goats in the Spider Silk Lab are regulated by the USDA, through the Animal Welfare Act. And because some of the goats are transgenic, the lab is also regulated by the FDA.[43] The lab complies with FDA regulations as outlined later in this section, although Justin does not believe the agency has ever inspected the lab. The lab also complies with the USDA, which does yearly inspections of South Farm to make sure there is clean water, adequate feed, and good and healthy animals.

Larisa Rudenko, currently a research affiliate at the Program on Emerging Technologies at the Massachusetts Institute of Technology (MIT), was a senior adviser for biotechnology at the FDA. Through her work at the administration, she was responsible for science-based policy and implementation for animal biotechnology through the Center for Veterinary Medicine. After Justin provided me with Larisa's contact information, I spoke with her to better understand the way the FDA works.

My conversation with Larisa was a little bit like getting a refresher in civics. She explained how a statute or a law passed by Congress becomes the overarching rubric that all agency's activities fall under. In this case, the FDA was created through the Federal Food, Drug and Cosmetics Act. The executive branch writes the set of regulations that become codified and have the force of law, although they aren't laws. Larisa explained the challenges of regulations:

> The agency has regulatory authority over animals' genomes that have been intentionally altered. We want to insure and provide the public with confidence that the agency knows what is going on and that they will not be put at risk during the investigational stage (as opposed to research), it's the development stage—end of R through D. Somehow you have to walk this line between safety and public confidence and not putting too many regulations that it becomes so burdensome that science can't move forward.

In 2008, the FDA issued the Guidance for Industry Number 187, "Regulation of Intentionally Altered Genomic DNA in Animals," and a drafted revision has been considered since 2017.[44] Guidances, in FDA-speak, represent the agency's best thoughts and are used to inform scientists of best practices.[45] Abiding by the guidance in the case of genetically modified animals is primarily to guarantee that no such animals enter the food supply.[46] We do not know what would happen if transgenic goats entered the food supply (or if goat milk containing spider silk protein was consumed by humans), and it is unlikely there will be clinical trials anytime soon to determine the effects of human consumption.

While the animal, in this case the spider goat, is not a drug, it contains an article that has the definition of a drug. As Larisa explained, "A substance that will treat, prevent, cure or mitigate a disease is a drug—an article that is intended to alter the structure or function of the body of man or animals." The introduction of DNA into the genome of the spider goat animal alters the structure or function of that animal. Larisa explained that the FDA works with the scientists in a "highly interactive conversation" to make sure they "keep these goats from mingling or escaping."

As part of FDA guidance, the spider silk herd at South Farm lives in isolation from other goats, and as an added measure to prevent any mixing, transgenic animals are composted after they die.

Spider Goats Becoming Old Hardware

Considering the discursive and material creation of spider goats as a collaboration of heterogeneous human and nonhuman actors, I'm reminded of the term *boundary object*.[47] A boundary object is a thing or concept that is useful to different actors in different settings to achieve different ends; the same object is plastic in its ability to be constructed as meaningful to different actors. Additionally boundary objects enable coordination and collaboration among actors, even if they are not defining the object in precisely the same way. For instance, depending on the actor in the collaboration of donor insemination, human sperm is a boundary object. It can be a sellable body product or a way to make some extra money (for the donor), a product enhanced by technical procedures to make it sellable as a commodity (for the sperm bank), and a product selected and purchased to achieve pregnancy (for the recipient).

Spider goats are also boundary objects differently understood by actors in the spider goat social world. For military agencies, they are *technologies* to produce scalable useful fibers. For herdsmen, they are *farm animals* to be cared for and milked. For scientists and engineers, they are *living factories* to be optimized for protein production. And of course, there is slippage between the categories of actors, whereas Justin is indeed a scientist but his experiences with animals enable him to toggle between seeing spider goats as *animal* and *live machine*. The way spider goats are produced in the very real material sense and in the very vivid imaginary sense by all the human actors and agencies imbues the goats with affordances. As described by psychologist James Gibson, affordances are properties of objects that enable them to function.[48] If we imagine goats as an object for human consumption, their udders have teats that enable them to be milked—humans can use their hands to pull on the teats. But beyond what is readily visually affordances of goats, there are also perceived affordances, human perceptions of what is possible.[49] Humans have designed spider goats to be imbued with affordances. Through human transgenic innovation, spider goats become an interface of goat and spider genetic and biological information and capacities. Milk containing spider silk protein is synthesized in the mammary glands of transgenic goats. The spider goat system generates flexible and strong fiber proteins that are then purified from goat milk. Spider goats are designed to be technologies that have obvious and intentional affordances engineered into them. However, even though these affordances are fabricated into the goat system, it is not fail-safe. Affordances are not guarantees.

During my second visit to the lab, it was clear to me that the goats were not producing the same quantity of protein as they had in previous years. This change, combined with the increased expense of housing, feeding, and caring for the goats, was forcing an evaluation of them as an effective and reliable biological system. While the lab has not yet perfected the use of *E. coli* to produce spider silk protein, the scientists believe they are close.

Until another system can do what the goats do, the lab is stuck relying on this expensive option, but it causes a lot of conversations. Justin explained to me that they can "certainly produce protein in bacteria" but that they cannot yet do it with enough volume or purity to replace the goats: "We need a system to produce these recombinant [genetic material formed by recombination] spider silk proteins that will ease the purification and also up the recoveries and, in essence, replacing the goats, because the goats are expensive, right?" He went on to explain that the goats are time-intensive and require "a minimum of two people that are out there daily, usually twice daily, milking" and that the cost of maintaining them continues to rise. Justin told me, "I think you started at something like twelve thousand dollars a year. And this is seven years ago. And now we're over thirty thousand a year for the care of fewer animals." I nod as he explains this—that it is definitely expensive—and I can certainly understand the troubling economics of paying more for goat care while getting less from goat milk production.

Justin then reined himself back in and noted that the Spider Silk Laboratory depends on the ability to provide spider silk protein scientists. He added emphatically, "But right now, it's the most defined system we've got because we can put many, many grams of this protein on the bench and generally have it there at all times for people to go look at it.

It's easy for people to go look at different applications. So yeah, they're just an expensive necessity."

Later in the day, I duck my head into Randy's office. "Got a minute?"

He welcomes me in with a wave. He is standing in front of his screen, peering at it. I recount my interview with Justin and a few other scientists in the lab and ask Randy, "So are the goats not viable anymore?"

Randy shakes his head. "Obviously, milk's viable because that's the best thing we have going right now, right? It's the most expensive but it's the best, you know we can produce more from that than we can from anything else."

I try again: "So would it be correct to say the goats are old hardware?"

He nods vigorously. "Absolutely. The goats are over."

I scrunch up my face, selfishly thinking of my book and how I missed the boat. "OK then. I guess I am having trouble understanding. It seems like there is a general sentiment that is different from my two-years-ago visit, where people want to phase out the goats, even though they are the most viable."

Randy quickly answers, "Oh, we would love to get rid of the goats. If they were still at the level of production that they were before, the answer is, we'd be perfectly happy to keep going because they're outproducing anything that we have. But they're not anymore."

"So why not just reproduce a new generation, and maybe they'd be better?" I ask, practically blurting it out, as if a new generation were that easy.

Randy's patience is so steady. "That's an interesting question," he says. "And I think part of it is that we got complacent. We didn't really follow each goat in terms of how much protein she was producing. And I think as a result of that, we haven't been pushing to truly identify the better producers and keep track of their offspring, even though, for reasons that we don't know, we did have two of the best goats, and they never threw a female; all they had were males. And then, you know, that's obviously useless for us. So, you know, I think that's part of it, is that we need to go back and really be much more careful with identifying the goats that are at the high-end [of] production and keep track of their offspring."

"So, how many more generations of goats do you think there will be?" I ask.

Randy shrugs and replies, "I got no idea. You kind of hate to end it. But you also can't afford to spend thirty thousand a year on goat care and on feed and all these kinds of things. Man, the price has been going up astronomically faster for reasons that we don't understand. They say feed costs. I mean, I know what the cost of feed is, and it hasn't gone up 50 percent in the last two years. It just hasn't."

Justin concurs, telling me, "I think our goats probably have a defined timeline here. What that timeline is, I don't know, depends on some of this other research. But because of the cost of it, you know, it's not necessarily advantageous to continue down that path. And you can imagine, you know, if somebody has a real interest, say, you know, real-small-volume, high-dollar-application spider silk goats, maybe it will satisfy that market. But anything that goats can do, we strongly feel alfalfa and bacteria can do better."

To be clear, Justin and Randy are scientists who are modifying biological systems (goats, *E. coli*, alfalfa, and silk worms) to get at a material, spider silk, to create products. Along with creating a good, reliable, and affordable system for producing the product, they also need to experiment and innovate with the product to demonstrate its utility to the rest of the world, particularly biotechnological companies that will mass-produce these inventions. However, Justin and Randy realize that having goats as the system is unlikely to work for anyone, because of the cost, size, and variability of goat production and reproduction. Over my time studying these goats, it has clearly become the twilight of their utility. Although they give the most protein of any biological system for reasons explained above, they are not a sustainable system. But what do we do with transgenic goats that are no longer useful to us?

The Cull List: Becoming Obsolete

The word *cull* comes from the Old French *cuillir*, "to collect or gather." The expression *to cull a herd* (not necessarily killing but to select the best animals) originated in Australia or New Zealand circa 1880s. The term's meaning, used in this book, in the sense of killing animals to thin out a herd or a population of wildlife, appears later, in the 1930s.[50] Making an animal killable is beautifully explored by Tara Mehrabi in her work on Alzheimer's disease research and *Drosophila* (fruit flies). Mehrabi describes the spectrum of killability: "Flies become killable in relation to other animals and in different laboratories as a matter of biological specificities, scientific credibility, and the technical and discursive scientific infrastructure built around them such as Drosophila genetic banks as well as their alienation from the category of animal."[51] This type of human management and discursive attribution of the goat determines when a goat is a good milker, a good reproducer, a transgenic protein system, or biological waste and thus killable.

Brianne "Bri" Bell is an undergraduate researcher who majored in biological engineering and has been working in the lab for eight years. For two summers, I spent some time in the lab just following Bri around and asking her questions about what she does and why. Much of her purification work is discussed in chapter 3, but Bri is also responsible for helping to decide which goats are placed on the cull list. During a routine Monday lab meeting, Bri, Amber (a herdsman), and Brittany Grob (an undergraduate biology student) stay behind after the larger group disperses to discuss the cull list. I scramble to keep up with their conversation. They are going through the herd and qualitatively evaluating their milk productivity.

Bri says, "[Goat] 718's protein is low. And 602 has a lot of fat content in her milk, so that is changing the way we have purified her, but she is pretty consistently good. And 607 is bad. We had to dump her milk"

Amber asks, "What about her, 623? She is up in the air. She was on the cull list."

Bri responds, "We're drying her all right now with all the ones that were on the cull list. We started drying them now. But she is decent—do we want to keep her culled?"

When I look at them confused, Brittany responds to my tilted head. "Culling them was like getting rid of them," she says. "You don't want them anymore." Drying off a goat means giving the goat a hormone to decrease their milk production. "When we want to cull them," Brittany says, "we want to dry them off so she doesn't get mastitis or anything. And sometimes we decided to keep her around for breeding in the next round." Drying off a goat can cause discomfort if it is done abruptly and without management, as an engorged udder creates pressure.

I left the meeting feeling disturbed. The manner of speaking about the cull list was so business as usual; it destabilized me by its ordinariness. I

Figure 2.5. One of two walk-in freezers with transgenic goat milk. Researcher Brianne Bell said, "I've got freezers full of milk, more milk than I know what to do with." Photo by Lisa Jean Moore.

was upset with myself for being judgmental of these people in my field site. They were not cavalier or unethical people. They were open and honest in front of me, patiently explaining every minuscule detail of their work. And still I couldn't shake how the clinical measurement of a goat's worth was the spider silk protein in her milk. How could I represent these people without my own judgments seeping out?

I approach Bri a day later and say, "I wanted to talk to you about the cull list. Why does an animal end up on the cull list?"

She sighs. "A couple of reasons," she says. "Usually because they don't produce a lot of milk or a lot of protein. If they do produce a lot of milk and they have a lot of kids and those kids need milk, we might keep them off the list, but we also usually have nontrans goats to feed them, so it depends."

Trying not to sound too defensive, I ask, "What is a lot of milk or protein?"

"At least probably a liter each time and at least a gram per liter (or a little over) per goat is enough protein. Anyone not doing that—we can't keep them."

"Bri, I am just wondering if you could tell me what does it feel like to put them on the cull list?"

"Oh, I still feel bad, every time," she says. She looks down, and it's clear she doesn't want to share any more about that.

I try to nod as sympathetically as I can, but I realize my questioning her about this issue invades her own emotional scaffolding. The emotional ecosystem is rich here—for the goats, the scientists, me. I modulate my voice; I recede, tack, come back another way. But in talking about culling, I am limited in managing competing emotions at the same time. I switch tracks and talk about purifications. The remainder of the interview is slightly stilted, and I'm distracted by the affective dissonance, so I wind down.

Later that day, I ran into Justin and tried to get more clarity on culling, asking him to explain what happens when a goat is culled. Justin said that there are "rules that are specifically designed to keep this transgene more or less out of the environment and to guarantee its breakdown so that it doesn't enter the food chain." In practice, this objective means that a vet comes and injects the culled goat with "a substance called beta-Euthanyl, and it provides a very quiet, calm passing," Justin told me. He said that the culling is "a really unfortunate circumstance where we do have to euthanize them; we try hard to involve other groups."

Even when they are killed, Justin noted, the goats continue to be useful:

> There are other groups here at the Utah State University that require tissues. So there's if you can give anything to make it more meaningful. Also we've got the start of a new veterinary school so we all involved the veterinary students. They can come practice necropsy, if a disease is present. They can look at the pathology of the disease, right there on site, put their hands on it. They can practice all these techniques with an animal.

As we continued to speak about culling, Justin added, "Mature cull non-transgenic does are 'recycled' and given to other projects out at the farm that does a lot of cloning. Those culled does' ovaries are harvested and used for cloning. We only cull immature does for humane reasons, i.e., the animal is sick or injured beyond fixing."

I asked Justin what happened to the bodies of the culled transgenic animals, because I was still confused by the various accounts of burning and burying that I have heard. He explained that they used to incinerate all the euthanized transgenic animals but that incineration was "absurdly expensive." So they worked out an agreement with the regulators to just "compost the animals because that causes the breakdown as well. So essentially, you know, they're buried."

To compost an animal such as a goat, there needs to be a precise carbon-to-nitrogen ratio that produces enough heat to break down the carcasses. According to animal geneticist Alison Van Eenennaam of the Department of Animal Science at University of California, Davis, "It is actually the best way to deal with organic matter since adding wood chips and shavings to the carcass creates good heat, and it becomes steamy as the microbes are chewing up everything. It's sterile and destroys harmful pathogens."

But ultimately, how will researchers decide that the goats are obsolete or too expensive? When is a goat a form of waste, and since federal regulations deem a transgenic goat's body dangerous waste, could this designation make it easier to kill the goat that was previously useful? Will the purifications be sustainable as a means of providing protein? How these decisions will be made is unclear. As Randy prepares to retire, he speculates about the end of the program: "I expect that Justin, USU, and UW would make the decision to stop using the goats. I am sure that they would just stop breeding them and wait for the natural process for them to die. The USU farm has a number of goats for other projects, so they could just be kept with the other goats. Unfortunately, they could not just be farmed out to someone to keep them until they passed away, as I am sure the USDA and FDA would not allow that, as logical as it seems."

I'm agitated when I think through this life course. The goat is born, typed for transgenic classification, and taken from her mother. She grows in confinement and is bred, and her kids are removed. She lactates, is milked, is freshened, and is bred again; the kids are again removed; and she lactates and is milked again—rinse and repeat until she's too old to do it anymore. Then she is culled and composted. These goats cannot be put out to pasture, euphemistically or actually. And even when they are surpassed as a technology, they won't get to live out their days in peace, having served the biomaterials industry.

3

The Goat As a System

From Pure to Artificial and Back Again

One way—the predominant medical way—to describe how I conceived my children is to say that I used *artificial* insemination.[1] This method of delivering semen to the cervix or the uterus is artificial (as opposed to "natural") because it does not involve penis-vagina sexual intercourse. Personally and professionally, working at a sperm bank in the mid-1990s; doing research for, and writing *Sperm Counts* between 2000 and 2007; and using frozen and fresh donor semen to become pregnant in 1997, 2000, and 2008, I've spent a lot of time grappling with this concept of artificial insemination. The word *artificial* carries a pejorative connotation—it can describe the personality trait of being insincere or a product that mimics an otherwise natural and superior version of the same thing. The adjective *artificial* implies that the penis is the legitimate, true, and pure device to get semen near the cervix and, moreover, that heterosexual vaginal intercourse is a natural activity. To use another method of seminal delivery such as a syringe is then phony or unnatural. Although female animals, including women and queer people, have conceived by means other than heterosexual intercourse for centuries, there is still stigma and discomfort associated with human-assisted reproductive technologies.[2] It's similar to the term *alternative family*, which came to prominence in the 1980s and is used to connote a family, often composed of same-sex parents, created through adoption, surrogacy, or artificial insemination. I cringe when I imagine overhearing someone say, referring to me, "She gave birth to three children using artificial insemination to create an alternative family."[3] Such a scene is not so far-fetched, since we've definitely been the poster family for many heterosexual progressive (and undermining) bragging rights.

Beyond my personal experiences, I also am fascinated as a sociologist by the complicated cultural and linguistic gymnastics at work here—phrasing that retains the primacy of phallic power, heteronormativity, and naturalness. The use of terms such as *artificial* and *alternative* serve ideological purposes better than they serve descriptive purposes. Those who cannot (or, more accurately, don't want to) conceive "naturally" must do so in another way and potentially use the tools of science to obtain donors' fresh semen or synthetic technosemen (washed semen from a sperm bank, described later in this chapter). All the while, we linguistically remind people that this form of reproduction is artificial, sullied by commerce, technologically mediated. The families they produce are variants that deviate from the essential, real, conventional family.

Sometimes, insemination requires the collected semen, a mixture of sperm cells and fluids (e.g., fructose and enzymes), to be washed or technologically prepared through protocols of diluting and centrifuging ejaculate. The resulting technosemen is a version of sperm that is synthesized through laboratory procedures to improve the odds of achieving conception. The washing of semen creates a concentration of sperm cells to enable more precise insemination directly into the uterus—a purified pellet of sperm reanimated with a lab-based liquid media.[4] Ironically, the technosemen is augmented and enhanced through purification procedures that make it more concentrated than ordinary ejaculate. In essence, technosemen is purer than natural ejaculate. Purity here is a bit of a moving target—where even those who engage in artificial means to achieve pregnancy can gain access to purified semen.

I've wrestled with these ideas in my fieldwork and my sociological interpretations; my work (and life) is caught up in the space between the artificial and the real, the clean and the tainted, the wholesome and the queer, the original and the boosted. Anthropologist Mary Douglas suggests that human cultural feelings about pollution (the soiling or tainting of material or symbolic space) and cleanliness serve to preserve the social order.[5] Impure objects that encroach on that which is considered pure (or culturally prescribed as such) are viewed as dangerous, indicating a transgression to the social system. In much the same way, the replacement of a penis (pure, original, natural) by a syringe (manufactured, unnatural) enables a transgressive act and facilitates reproduction without a man present. Nonheterosexually coupled people are deviating from heteronormative reproductive scripts, thus subverting the social order. Their actions are then labeled as bogus. Scientific interventions may improve on nature and facilitate social changes (within reason). Over the past several decades, humans are creating a new social order pertaining to the nuclear, heteronormative family, but some vestiges of seminal power (however disembodied and souped up by capitalist investments) are also retained.

In this chapter, I describe a different process of purification—the extraction of spider silk from milk produced by transgenic goats. As in the case of technosemen, a sociotechnological collaboration of scientists, engineers, and animal bodies enables access to a biologically extracted commodity, pure spider silk protein. Purification is a scientific mediation and technical operation—goat bodies are modified by genetic donation from a spider, goats are bred to kid, and then transgenic milk is produced, collected, and filtered to extract the spider silk protein. In a straightforward operation, the transgenic lactating goats are milked in the barn, the milk is measured in a series of buckets, and statistics are recorded about volumes, dates, and goats. The milk is then transferred into huge zippered plastic bags and labeled. The bags are placed in plastic tubs, and the transgenic spider goat milk is transported via car or truck to the laboratory's freezer, where eventually a lengthy, multistep process of milk filtering begins. The spider silk protein, extracted in powder form, is then available for scientists to use in experimenting to innovate products (see chapter 4).

Simply put, artificial animals, the spider goats, are constructed and put to work, making new objects and relationships. Similar to the case of artificial insemination and sperm washing, spider silk purification represents the replacement of an old social order (one that presumes animals are pure and natural) with a new social order (one that presumes animals are hybrids and transgenic). Science and the marketplace usher in a new social order; transgenic goats are systems (living factories) producing raw materials for human manufacturers. Our cultural disgust at the boundary crossing represented by the tainted transgenic animal is partly cleansed through the scientific process of making pure spider silk as a material and economic opportunity.

As I write about the making of the transgenic spider goats, I am also examining my own spliced-parenting fantasies, my wistful sentimentalizing of Darwinian exploration, and my sociological training to produce sound theory. Even in my methodological choices, I am pushing for a "contaminated critique," as anthropologist Katie Stewart urges, where instead of remaining disengaged from the objects I study, I am swept up with them—we are all contaminated and I am implicated by them.[6] I am questioning my own epistemological assumptions by disrupting the purity of the scientific binary subject-object relationship. Through my own entanglement with the ideas, I hope to shake up these binaries and develop multidimensional perspectives based on the relativity of the participants. As Stewart suggests, ethnographers and the objects they study do not form tidy relationships; rather, the relationships are messy and the yearning for understanding is filled with uncertainty, confusion, and conflict. Remaining pure is impossible and not preferable, but there is still this ideological drive to get to purity through my own observations (the "real story") and, in the Spider Silk Lab, the "purest spider silk."

My own genetic relationships are interwoven into this book—my relationships with my own children, with seeing myself as a reproductive organism on a continuum with goats, and with spiders. In this chapter, I begin with the purification, tracing the multistep process through which scientists turn the lactated milk of spider goats into isolated spider silk protein. I detail both the treatment of the milk and the scientists and their tools. In the rare moments where I find the unmeasurable variables at work in the process, I highlight them and how the scientists engage with them. I also note that, through the purification process, the lab is also producing new narratives about our relationship with other animals and their material products. I then turn to a brief history of synthetic biology, especially how the socioeconomic environment pressures science to be increasingly profitable, effectively structuring the conditions under which science is conceived and accomplished, and what this means for the spider goats and the scientists who work with them. Finally, I examine how scientific systemization, quantification, and alienation have turned these goats into a device or an apparatus to augment us. This is not to suggest that once-pure goats have been contaminated or made impure by the science of transgenics. Rather, while previous human-goat interactions involved a co-constituted history of entanglements, goats have now become useful to humans not as goats. The goats are now a system of production, making something entirely not of the goat but the spider silk protein desired by scientists (and the market).

Pure Nostalgia

I've long had a very vexed relationship with Charles Darwin. His fragility and tenderness appeal to me. I have a soft spot for his lifelong vulnerability to a weak belly, painful gut, and endless days of vomiting. His love of his children and Emma, his wife, move me. I read his letters, biographies, other accounts.[7] In a letter to Emma before they are married, he refers to his stomach and anticipates the pleasures of being with Emma in marriage over the professional tasks of grand theorizing: "I have no very particular news to tell you, as you will guess by my having written so full an account of my stomachic disasters. . . . I think you will humanize me, & soon teach me there is greater happiness, than building theories, & accumulating facts in silence & solitude."[8]

So ordinary in his extraordinariness. As a scholar, I admire his skills of observation and concentration and his many talents of dissection, description, and drawing. Yet, as is usually the case with intellectual crushes from the nineteenth century, Darwin is a complicated and flawed figure. While I sympathize with his interior and marital conflicts over religious beliefs, I am also disgusted by his racism and sexism. Even with some nascent humanitarian and abolitionist ideas, his work was undeniably inspiration for eugenic and imperialist thinkers, practices, and policies.

Sometimes, even still, when I struggle with insomnia, I imagine the world through his eyes. Lush and colorful, a proliferation of life, bursting, free and pure of human tinkering, corruption, and interference. I have a certain old-world nostalgia for Darwin (like my *Little House on the Prairie* fantasies of Utah)—I dream about traveling on a boat for five years. But nostalgia is a sticky trap, swept away by my urgent longing for a simpler past that never was. Me in nineteenth-century attire, wiping sea spray from my face with a handkerchief, nodding as Charles describes a mollusk he turns over in his hand. Maybe it's weird that a queer middle-aged feminist in New York City is so consumed by anachronistic abstractions and impossible romantic sequences. But I indulge myself and construct an earth before airplanes, the internet, cruise ships, invasive species, and the Anthropocene. I dream about being Darwin's sidekick on the *Beagle*, spying the signs of the flourishing Galapagos. Bright sunshine, blue-green oceans teeming with sea creatures, white sails balancing on deck as we precisely calibrate the dials on mechanical instruments or sketch the dolphins flanking the boat.

I project all this purity, innocence, and lack of moral gray areas into my fantasies. It is not lost on me that these fantasies are based on dominant phallocentric tropes of man conquering nature, but my life subverts these tropes.

Waking up from such lucid dreaming on a chilly Brooklyn November morning, I hustle Greta to the bus stop as she complains in great detail how her cowlicks are messing up her haircut. I nod in absent-minded maternal sympathy and whisper, "Whoa, a bad-hair day at ten is rough." She frowns, sensing sarcasm. The crossing guard hollers her customary, "Mornin', Pumpkin." Thankfully the bus comes on time and I slip into a nearby café.

In the café, National Public Radio is playing Trump's first impeachment hearings, and I wince but decide to stay and grapple with a pile of textbooks about synthetic biology. Quickly frustrated and, frankly, bored, I text my oldest daughter, Grace. She's working on her college senior project at an on-campus lab, a study of what makes what makes some strains of *E. coli* uropathogenic (causing disease in the urinary tract). I am hoping she can help me make this textbook information more interesting and understandable.

> LISA JEAN MOORE: Two questions: What is the relationship between DNA and protein? And why is it difficult to directly synthesize a protein from chemicals? Why do you need a biological system? Like a goat or E. coli?
>
> GRACE MOORE: DNA gets transcribed into RNA, which gets translated into proteins. So the protein-coding regions of DNA, when activated, get turned into proteins. The chemical synthesis, I am not as sure about, but I imagine it is because it is hard to do the translation step (from RNA to protein) without ribosomes, because as far as I know, no one has made ribosomes outside of a living organism. The ribosomes are the point at which nucleic acids are translated into amino acids.
>
> LJM: So the book I am reading says you need a cell to make proteins.
>
> GM: Yes, all cells have ribosomes. And protein folding is really complicated.

Being Grace's virtual sidekick as I brave it out in Brooklyn is nowhere near as fanciful as my Darwinian excursions. And I am not sure I really get it. When I read things like "folding protein is really complicated," I imagine an intricate origami animal and try to work my way back to the work of ribosomes with their tiny hands holding little square pieces of colorful paper. Protein cranes soar around in my head. This experience is so disembodied as I text her from a café without the energy I experience with Darwin. I continue trying to generate the somatic thrill I felt in Florida—I want that feeling of leaping through the tangled jungle and holding spiders.

> LJM: When goats were genetically modified, their DNA was changed to add a gene for spider silk production and then the RNA in the ribosomes went to work to make the spider silk protein, which was expressed in the milk?
> GM: The RNA isn't in the ribosomes; it gets translated by the ribosomes. But yes, ribosomes are their own thing.
> LJM: I'm sorry to be so dumb. But I guess what I want to understand is, you change the goat by adding this gene or DNA (?) to make it so the RNA makes/expresses spider silk proteins? This is all really helpful, by the way.
> GM: You're not dumb! It's complicated! Yes, you insert the gene into the goats' DNA and then that gets transcribed into RNA using the goats' normal transcription methods and then the RNA gets translated into an amino acid chain which folds into the protein. The idea is that if the gene is successfully inserted, then the goat's cell will do the work for you in the normal way that all the goats' genes are expressed (transcribed and translated).
> LJM: I appreciate how smart you are. Go back to having a life.
> GM: Literally waiting for E. coli to grow.

There is something very flat and dull about this interaction, despite my imagination in overdrive. I feel washed out. I've had a hard time writing this chapter of the book, confronting this dull feeling. But have I constructed this binary—this stark contrast between Darwin's science (before, original, pure, rich, dazzling) and Grace's (after, derivative, impure, tedious)? Darwin created a theory of evolution that both confirmed the connection between living animals and insisted on the distinction between species; he argued that mutation happened over long periods and in response to the environment. His theories are built on detailed narrative descriptions based on drawings, sample collection, dissections, measurements, and observations. He used various modalities to richly describe biological organisms. One of the reasons why the ribosomes are so hard for me to understand is that they're hard to visualize. Here I am, centuries after Darwin, studying the field of transgenics, a field that has collapsed time and created mutation and in which I am inserting description. I'm trying to work out the connection and distinction between living animals in another way.

My work here and in my previous book on horseshoe crabs treads through some of these same themes of enchantment and disenchantment, or wonder and routine. My fieldwork often juxtaposes how I hold in my consciousness the sense of enchantment, wonder, and artistry of biological discovery while also recording skilled technicians breaking biology down to knowable and mechanical parts through so-called advancements. As I have previously admitted, it is bad colonial anthropology to create such binaries of then and now, of before and after, of pre- and postcontact, industrialization, Anthropocene, steam engine, imperialism, Darwin. And I have been schooled by contemporary social theorists to know that thinking of time before the Anthropocene is wrong—it screws up our ability to live in the now, to deal with what is happening right this moment.[9]

In fact, I remember a precise moment of my own consciousness-raising on this issue, this idea of glorifying the nonexistent before. I was taking a cultural studies course with Jim Clifford at the University of California, Santa Cruz, and we were reviewing the work of Raymond Williams's book *The Country and the City*.[10] Clifford's explanation of Williams's use of the bucolic and the pastoral to reveal the trap of nostalgia honestly blew my mind. The past, the country, the unvarnished natural landscape of the preindustrialized is conjured to pretend as if all life were void of machinations of power. This vision enables us to prop up some notions of innocence, shirk our responsibility to our legacies of domination, and suggest that some time existed before this industrial contamination of Eden. These visions of pure nature mobilize us to reflect back to a time that never was. It exacerbates our penchant for the dichotomy. I also see that this drive in sociology to categorize can mean we fall into the binary about enchantment and disenchantment—rather than the messiness of lived experience. This sociological imperative can prevent me from perceiving the robustness of more-than-human life.

It doesn't take some fantasy about the untainted natural to feel transcendence. I have experienced the miraculous electricity of discovery in the most mundane of places (the side of the road, the shores of a filthy urban beach, forgotten Brooklyn rooftops).

There are complexities—a narrative messiness that does not neatly fit into traditional dominant narratives about the distinction between the natural and the unnatural or about the pure and the contaminated. In fact, purification automatically conjures up the impure. Something is qualitatively different about engineering a new species in the lab to do things, to make raw materials, to make other things.[11] *The animal has become a machine.* Plus, I've been trained to create an argument—after all, that's how you get a book contract—but if I attempt to fit things neatly into some dichotomy, I could miss the nuances of lived experiences I so desperately want to capture. My internal dialogue is a constant chatter of "Are the spider goats bad? Of course they are, and why is this bad? . . . Capitalism, duh. But isn't that too simple? Can't the goats be enchanted and magical too?" I want them to be more than just bad.

The deterministic superrationalization of late capitalism has worn down our capacity to wonder.[12] It is easy for me to slip into an automatic, critical sociological drive to expose this evil apparatus that flattens the human spirit and alienates us from our species being.

I don't know precisely how, but I'm sure Grace's research is connected to Darwin's, and she is indebted to him. Whereas Darwin and my imaginings about him feel effervescent, what I learn from Grace is highly specialized knowledge analogized through words like *translation*. It's official, legitimate knowledge that she learned through textbooks and PowerPoint slides not through experience, observation, or research. But the science I have gleaned from Darwin's accounts entices me to observe as much as possible and to be careful and judicious in the act of creating a taxonomy. His scientific method was interdisciplinary and expansive, painstaking, and slow but was also conducted in the field.

I'm struck by the rote plodding lab work that is required of Grace and of the scientists I've met in the spider silk purification lab. They repeat time-consuming, multistep procedures in sterile, built environments under fluorescent lights. They're workers implementing a set of directives more than they are scientists in the Darwinian conceptualization. Advancements in synthetic biology appear to offer evidence of scientific progress, but as I'll explore later in this chapter, I question my own assumptions about whether these advancements take away from the magic, the wildness, the unpredictable, the irrational. My fieldwork has led me away from clear-cut, certain theoretical conclusions and has required me to make room for more ambivalence and contradiction in my conceptualizations.

Purifications

Purifying transgenic goat milk is a multistage process beginning with milking the goats, as described in chapter 2. Goat milk is transported by truck from South Farm to the Spider Silk Laboratory's walk-in freezers five miles away. There it is kept in large plastic bins until the lab is ready to thaw the milk from a particular goat to retrieve the specific protein. From the freezers, the milk is thawed in a warm water bath in a lab sink. Milk from Saanen goats is typically between 2.9 and 4.0 percent fat.[13] Since this fat clogs the purification machine, thawed goat milk must first be defatted before it is purified. When the purification machine clogs, it must undergo extensive cleaning, which is "a real pain," said Bri Bell. Thus, the defatting step is important to the scientists. The cream separator, named Milky, is therefore an essential piece of lab equipment. Purchased for around six thousand dollars from a company called MilkyDay and based in the Czech Republic, it's made of stainless steel and aluminum and works as a centrifuge. The skim milk is separated out to the edges of the spinning bowl and is siphoned off through an opening, and the heavier cream (the fat) sits in the center of the bowl and goes out a different opening. The smooth, fatty cream flows down a foot-long metal chute into a plastic container in a hypnotic steady stream. This creamy bucket brings to mind delicious dairy treats. The other metal chute collects the skim milk, which runs faster to a separate collection bucket.

Figure 3.1. Defatting milk with the larger machine called Milky. The fat is separated and collected in the container on the left, while the milk is collected in the larger container on the right for purification. The fat is dumped down the drain, and the milk is taken to the purification machine. Photo by Lisa Jean Moore.

Between five and forty liters of milk are purified at a time, depending on the type of protein (MaSp1 or MaSp2) being purified.[14] Walking into the middle of the laboratory, I hear the familiar rattle and hum of the purification machine processing goat milk through its tubes and filters. It sounds not unlike a train on the tracks, with rhythmic chugging and clicking. Approaching the machine, I have to suppress a smile, as it always strikes me as funny to see the DIY Tinkertoy appearance of the purifying contraption. It looks more like a science project created by the winning teen at a local science fair than the centerpiece of a lab extracting transgenic spider silk for medical and industrial use. I had expected a sleek Japanese-designed, German-engineered product with shiny buttons and monochromatic steel siding. Instead, its guts are on display, and when in action, the gauges and piling slightly vibrate. The whole creation reminds me of a miniature and mesmerizing Rube Goldberg machine with three lab technicians starting and checking on the progress of the filtration over two days. This contrast speaks to the realness of science in action versus the fantasized sci-fi aesthetic we have been socialized to expect. The scientific imaginary can be so much more satisfying than the actual doing of science, with its mundane, repetitive labor of collective action and tedious tasks of measurement and classification.[15]

The lab started purifying goat milk in 2003 at the University of Wyoming. Over the years, a lot of tinkering has taken place in pursuit of "upping the recoveries," as Justin put it, or maximizing the amount of spider silk protein purified from the milk.[16] I asked Justin why the word *purification* is used for this process: "Isn't it just really an extraction of what you put in the milk in the first place?"

Justin answered, "The process is a purification, as we are isolating our target protein away from all of the contaminating milk proteins. The process of purification does begin with filtering the milk at the molecular level. We use the size of the fat micelles and normal milk proteins as a means by which to effect the first purification. We then selectively precipitate our spider silk protein away from the other milk proteins as a final step to purify the protein to greater than 80 percent purity." In this definition then, the goat milk has become a contaminant that must be drawn away from the spider silk protein, or the target. The milk runs through the filter columns thousands of times over the course of a purification run.

And still the irony of the term *purification* astounds me. Pure means unaltered, unsullied, how it came, the way it was intended. As an adjective, it seems that *pure* is complimentary, indicating a feature that is socially desirable. Milk is a signifier of purity, innocence, youth, fertility, nourishment, and abundance.[17] In the United States, capitalizing and leveraging this symbolism produces solid market behaviors and is culturally propped up by a partial reliance on racialized histories and white privilege.[18] In the case of milk's rise as a staple of the American's diet, sociologist E. Melanie Dupuis's history of "nature's perfect food" chronicles how cow milk became a first and indispensable American food. Her work traces how the production and consumption of milk is as much a product of cultural ideas as it is of materials needs; herein also lies a history of whiteness.[19] Human consumers appear to be very persuaded by a discursive classification of things, especially food, as pure, with labels of organic or natural.[20] But just as cow's milk is an adulterated product, spider goat milk is not pure.[21]

In the lab, spider goat milk is made to produce something pure, desired, and wanted—but here the milk itself is pollution. But somehow, the use of the term *purification* obscures what is actually happening. In the treatment of spider goat milk, purification is turned on its head, in this case leaving the foreign substance only. But for me it is as if they are throwing away all the real and authentic material that is the purest expression of the goat.

Goat milk, the very fount of multispecies sustenance—the pure, the animal, the natural—has been altered by human genetic engineering. The addition of the spider silk protein now means that the milk itself is soiled. But interestingly, goat milk is not considered undesirable debris to the herdsmen who collect it. They want to maximize its production, collect every last drop. Not until the milk reaches the lab and is processed for purification does the milk itself switch to the waste product. The endless flow of goat milk through tubes and other devices—from milk ducts inside the goats' body to the udders, through the milking machines at South Farm to the bucket, through the defatting machine to a smaller bucket, through the purification machine to the centrifuge to the smaller and smaller containers—each stage takes away a little bit more of the goat part of the milk and gets closer to the spider part. How does science demystify things—disenchant goat milk by making it garbage—but also how can science not make sense? The milk is, as Justin says, the contaminant and must be precisely removed so it does not interfere with the spider silk protein. The residual milk, after the spider silk protein has been extracted, is then dumped down the drain as an unwanted by-product. This precision, this extraction, this reduction, is purification.

Significantly, goat milk, which had been the vehicle to get to the precious spider silk protein, is no longer seen as an effective instrument operationalized for productivity. The process of purification means that goat milk and, by extension, goats are recast as waste, interfering with procuring the desired prize, the silk protein. After the protein extraction, goat milk is now a liability that must be managed. This recharacterization reminds me of when my friend was training to be an ob-gyn and she had the revelation that when a woman is in labor, she and her body become "the enemy, in the way of getting to the baby. Everything the mom's body does gets in the way, and you need to get that baby out as quick as possible in spite of the mom." Her observation takes the old adage and turns it around: it is OK to throw the mother out with the bathwater. With spider silk goats, the narrative is constructed like this:

- What was once considered pure, the goat, is made impure (transgenic).
- Impure material, the milk, is removed from the impure creature.
- The raw material resource, the spider silk protein, is extracted and made pure.
- The goat milk and, eventually, the goat herself, once constructed as pure, are now waste material.

As mentioned in chapter 2, Brianne Bell has been helping with milk purifications at the lab for at least eight years. During my fieldwork, I probably spent the most time following Bri through her work of purifying goat milk. She was soft-spoken, kind, and circumspect when she answered a question; she answered thoughtfully to be as accurate as possible.

I asked her what she thought was her particular skill set that enabled her to run purifications.

She sighed deeply and said, "It takes a lot of patience."

I laughed. "Couldn't agree more."

Bri explained the various permutations of the protein recovery system. Despite all the refinements in the mechanics of the purification machine and the procedures of handling the milk, the amount of protein can also depend on the "the type of protein purified and oftentimes also depends on the goat being processed. Traditionally, we could get eight to twenty grams of MaSp1 from one twenty-hour process, but lately we get around one or two grams because the goats aren't as good or have declined in production with age. The MaSp2 protein used to give us only one gram from a single ten-liter twenty-hour run. We now get between one and fifteen grams of protein, depending on the goat."

Realizing the complexity of her explanations especially to a layperson, like me, Bri apologized for "how confusing this is." I was struck by the next thing she added, almost as an afterthought: "There are so many variables in the process and the results, it sometimes feels like the weather or mood of the lab affects the process." The weather? The mood of the lab? I perked up. Bri was intuiting that the ephemeral, random, uncontrollable aspects affect scientific results.[22] I've come to yearn for this acknowledgment of unmeasurable variables and how we just sort

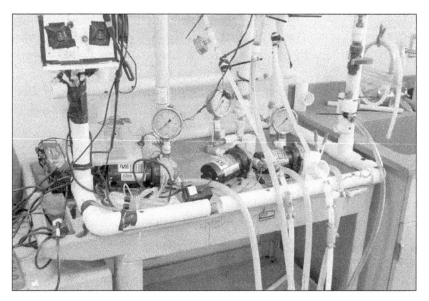

Figure 3.2. The purification machine designed and built by Randy Lewis, Justin Jones, and many others. This machine filters the milk thousands of times over twenty hours. Photo by Lisa Jean Moore.

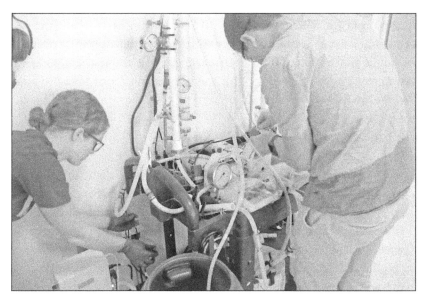

Figure 3.3. Two lab technicians prepare to do a milk run, or a purification of goat milk. Photo by Lisa Jean Moore.

of feel their effects. Are moods contaminants in the lab just as dogs are contaminants on the farm, both affecting the delicate calibration of science in action? As Bri was called away to attend to some delivery business, I pondered the idea of moods floating through South Farm and throughout the lab. Goat milk production is affected by how safe they feel, and scientists' purification processes are influenced by moods, pervasive attitudes, tones, or affective atmospheres.

At times, the scientists seem fairly consistent on the surface, exhibiting systematic blindness to the goat as an animal with emotions and lived experiences (beyond the scientific). But so much of the narrative generated during my interviews suggests an underlying personal struggle with the goats as sentient creatures and a professional need to think of them as machines: patented, owned intellectual property and machines. As Bri was indicating here, the lab itself is also an ecology of sensory ambiances and technological applications working in a multispecies refinery.

I asked Bri, "Why don't you just talk me through what a normal purification process would be?"

She shrugged. "OK. So frozen milk comes out of the freezer. We thaw it and then defat it. And then we'll treat it with our arginine and get the right pH for whichever protein we're working with, and then we load it onto size-filtration columns. Then we let that run pretty much overnight and watching the pressure is right. We are working on fine-tuning to make it optimal for the maximum amount of pressure and flow rate so that the greatest amount of protein goes through. We process it overnight."

Bri also described the machine:

> We use these filter straws to separate things out. It will go in through here, and then everything that is smaller than protein—or smaller than [the microscopic pores in] these membranes—[will flow through the membrane]. This is a point two micron [0.2 micron] filter. So everything that's smaller will be forced through our membrane, come out to this permeate line. This will include our protein, and it will also include milk proteins, so it will go into a collection bin over here. And then that will be sent up through this guy, which is only fifty kilodaltons in size and [this] protein

is sixty-five kilodaltons. Then it goes through one that is thirty kilodaltons. So everything smaller than thirty, which is pretty much all milk proteins, will then be forced through, and it'll come shooting through this permeate line, go through here, and that will actually be sent back to the milk line, which will be over here.

Bri was pointing to different tubes and collection containers as I took photographs and scribbled notes. "And that will then just continue flowing," she said. "So we'll have it set up to cross-feed, so then I can run it continuously. Filters aren't efficient when you just send them through once. They're more efficient if you send them through several times."[23] All the while, I couldn't help thinking that this entire machine did not look capable of this highly specific task. Instead it reminded me of something my brother and his friends might have made in the backyard in grade school.

When the researchers were talking about purification, I heard the term *kilodalton* stated in a way that was intended to impress me. Such as, "We're talking thirty kilodaltons," with an expectant look as I nodded as if I knew what that meant. A kilodaltons is one thousand daltons, and daltons are the units used to describe the sizes of proteins. Daltons are the equivalent to atomic mass units, or the masses of individual atoms of a given element. I asked Bri to explain it to me.

"Kilodaltons are a difficult thing to picture," she said, "but I'll try my best. Typically, proteins follow a rule of 1,000 kilodaltons being roughly the same linear size as 0.1 microns (100 nanometers, or 1×10^{-5} centimeters), and 50 kilodaltons being roughly equal to 0.004 microns, or 40 nanometers, or 4×10^{-7} centimeters."

This description left me even more confused. I turned to Grace for an order-of-magnitude estimate, calling her at the lab. She put me on speaker so her friend Maggie Prentice could assist. We discussed kilodaltons.

"Grace," I asked, "what can the human eye see that we can use to explain how small spider silk protein is?"

After doing some calculations, she told me, "It would take approximately 11,538 spider silk proteins to make the width of a human hair."[24] That's my girl. Grace has become a scientist, and she is engaged in the thrill and mundanity of routinized science. Yet she herself is the product of cutting-edge science. I wonder, is she purifying herself, making herself more mundane through this participation?

After the fluid was separated out, Bri performed a procedure called an ammonium sulfate precipitation. At the conclusion of the twenty-hour purification, there are two buckets: a bucket of a milk mixture, which has now been reduced to water and milk proteins, and a final bucket containing spider silk protein solution. Brianne then explained, "We remove our spider silk protein solution from the filtration machine and add enough ammonium sulfate to it so that our protein precipitates out of solution with the help of the forces from centrifuging the solution." We walked into a small room off the open laboratory, where an upright container was being swirled around and around. Bri pointed to this container and said, "We add enough ammonium sulfate to make the spider silk protein less soluble in water, making it easier to leave solution when combined with centrifugal forces. After centrifuging, the spider silk protein is out of solution as a solid pellet, while the liquid portion from the centrifuge (the supernatant) is essentially just ammonium sulfate and water." She continued:

> There is still some ammonium sulfate and potentially other salts from the milk present in our spider silk pellet, so we then have to break up the pellet and wash it with water or low amounts of isopropanol. This will release trapped salts into solution. And with further centrifugation, the salts are again kept in the supernatant and the spider silk stays in the pellet form. We repeat this process until the conductivity of the supernatant is below twenty microsiemens per centimeter [20 μS/cm], an indication that we have removed as much salt as possible.

Holding up a small glass capsule containing a white claylike substance, Bri said, "This is what the resulting substance looks like. The two proteins are indistinguishable, so we have to carefully label them." The stuff looked nothing like the golden webs billowing and glistening in the Gainesville sunshine. Watching Bri work through a purification as a practice and listening to her describe purification as discourse, I began to wonder. Instead of the monolithic characterization of science as a metanarrative disenchanting entire cultures, perhaps there is an ongoing

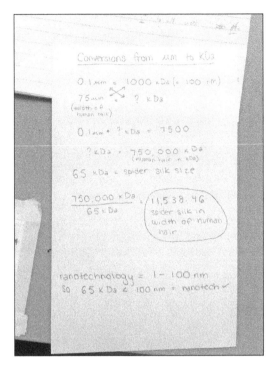

Figure 3.4. Grace Moore and Maggie Prentice's calculations to compare the size of spider silk protein with the width of a human hair.

process of enchantment and disenchantment as new discoveries disrupt the taken-for-granted assumptions of how the world works. The work done in the lab is not just the purification of material; it is the development of new narratives that construct our relationships with other animals and other materials in relationship to the means of production to legitimize these means.

It is difficult for me to imagine how all this labor, decades of spider silking, the raising of livestock, selective breeding, the calculations, various innovations, and tinkering with machines produced this small vial of white goo. This purification process yields a small amount of spider silk to be used in the products that scientists are hoping to make and are currently making. Before analyzing the products being proposed and

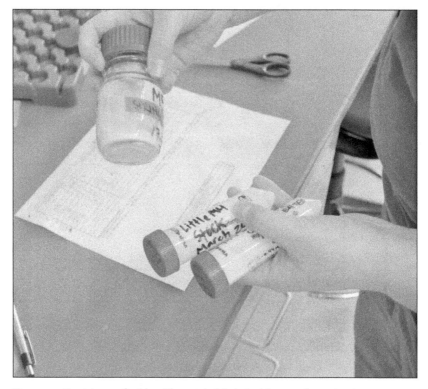

Figure 3.5. Containers of spider silk protein labeled with type of protein, date, goat. Photo by Lisa Jean Moore.

created, I want to examine this production in the context of synthetic biology and the pressures that scientists face in a market-driven field of knowledge production.

Brief History of Synthetic Biology

There is a time-travel quality to synthetic biology, where we can invent a new common ancestor of the goat and the spider. Breaking the rigidity of the species barrier is repeatedly celebrated among synthetic biology accounts.[25] Geneticist Adam Rutherford describes synthetic biology as a new way of creating life without the limits of species-specific reproduction. "With synthetic biology, we can now bypass sex and breeding altogether and put together organisms not even slightly capable of having sex, having been separated on the evolutionary tree by hundreds of millions of years—spiders and yeast."[26] And while we can "put together organisms" in the case of spider goats, sexual reproduction is certainly necessary to continue extracting spider silk protein from the milk and to create new generations of spider goats. Writing about spider goats, Rutherford considers them a chimera, where "genes have moved across the species barrier."[27] He also explains, "The last common ancestor of a spider and a goat would have existed something like seven hundred million years ago, at a time when the beings that would acquire hard shells, such as insects and crustaceans, were evolving away from creatures with fleshy exteriors, such as the fishy or reptilian beasts that would eventually lead to us."[28] But this engineered chimeric species is made to do something, to make something, to generate a piece of technology with yet-unrealized potential.

With its roots in biomimicry (the practice of emulating and imitating nature's forms, patterns, and practices to solve human problems), synthetic biology attempts to manufacture products through biological systems. Since about 2003, synthetic biology emerged as an interdisciplinary field that applies engineering principles to biology with the objective of redesigning and fabricating biological components and systems that do not exist in the natural world.[29] The fields of genomics, chemistry, and mechanical engineering are combined to manufacture cataloged DNA sequences and create new genomes. The funders of synthetic biology are mostly interested in clean energy, bioweapons, and cheap drug synthesis. This type of data-driven science means there is less interest in asking questions than there is in finding correlations and knowing things (through data that is gatherable). Data that is harder to gather or that is not as purposeful is consequently ignored.

Ending extinction for long-dead species, adapting humans for life in outer space, and eliminating death, to name but a few, are the key goals of this field of biology.[30] Yet there are more modest goals of synthetic biology, or at least there are building blocks, such as silk milk, where we can trace how larger claims are mounted. Spider silk protein from goats—silk milk—is one expression of synthetic biology, and as an object, an actant, a network, and a node, silk milk draws in different players and networks and reproduces certain forms of life. These forms serve certain interests, scientific and military and biomedical.

Cultural anthropologist Sophia Roosth spent eight years working with synthetic biologists, both interpreting the emergence of the field and tracing its recent fifteen-year history. Her work explores how "engineered organisms smuggle conventional ideas about creation, kinship, property, labor, democracy, and species beneath their immaculately ahistorical membranes."[31] Her observations, primarily at laboratories of MIT, revealed how synthetic biologists see themselves as constructing, shaping, modifying, or designing but never creating life; these scientists were clear that they didn't want to get embroiled in any sense of religious connotations of creation. Rather, some synthetic biologists construct "transgenic critters" that rupture the notion of species as the scientific investigations "amalgamate genetic material found in distant branches of the phylogenetic tree."[32]

Specifically, at my field site at USU, spider silk is a signature product showcased by the Utah Science Technology and Research (USTAR) Synthetic Biomanufacturing Institute. To fund the bench science, including the development of applications for spider silk, USU created the USTAR initiative in 2006. As described in the introduction, USTAR is an entrepreneurial venture to develop incubation facilities for academic and business partnerships. Its mission is "to accelerate the commercialization of science and technology ideas generated from the private sector, entrepreneurial and university researchers in order to positively elevate tax revenue, employment and corporate retention in the State of Utah." In addition to material production, the creation of spider silk and the systematization of its production is a semiotic frame.

The work of the Synthetic Biomanufacturing Institute, including transgenic goats, is at once about this redesign and fabrication of biological components and systems that do not exist in the natural world but also more specifically about monetizing such efforts. Anxieties about what it means to alter nature, even as the promise of things like ending extinction and extending human life sound compelling, are compounded when these alterations are motivated by more than the pureness of science. Under neoliberalism, science will always need to have an economic end to justify itself. Therefore, science for the sake of pure science seems impossible. When science that already exists at the frontiers of uncrossable boundaries is transparently a capitalist pursuit, discomforting images of the evil scientist abound. Since the turn of the new century, universities have become more entrepreneurial with the growth of science parks, incubators, and technology licensing arrangements.[33] There is concern about the maintenance of academic integrity and ethical scientific practices with the advent of venture capital in the university setting.[34] Many worry that taking shortcuts to produce results brings shoddy or dangerous innovations to market or that corporate interests set agendas for scientific research according to profit margins rather than utilitarian interests for the public.

While in Utah, I met with Christian Iverson, the director of Technology Transfer Services at USU, who specializes in commercializing life sciences technologies. We had a lively conversation about how he works to establish partnerships with companies interested in bringing spider silk innovation to market. As he explained his role, I imagined Christian as a PR bridge between the scientists and the corporate interests or venture capitalists. Christian described the USU scientists as "very dedicated to the science and not market driven." It felt almost as if he were protective of the reputation of the scientists as distinct from the corrupting influences of the market. While I observed this scientific dedication to be empirically true, my conversations with Randy and Justin revealed how they are keenly aware of the pressures to make spider silk commercially attractive and viable.

Furthermore, Christian explained, there are clear procedures and bureaucratically established norms for how invention is "owned" at USU.[35] "The intellectual property is patented in the inventor's name," he said. "So, for instance, with Randy and Justin, these patents were issued, and they're listed as the inventor. Part of their employment contract is that they will assign those intellectual property pieces to Utah State University."

It sounded strange to me. A concept that I have coined, like *technosemen*, is not assigned over to my university system—but again, my term is an intangible idea that has never generated any material product or physical commodity.[36] Befuddled, I asked Christian to explain who owns the patent.

He said, "So what happens is we preview the disclosure and we review the invention. If we feel it has potential, we feel it is important to the university, then the university will invest in that and file for patent protection. My office will work on the commercialization side, reaching out to companies and finding and [negotiating] deals and those types of things. And then . . . a hopeful result is that part of our commercialization will receive revenues back into the university. Those revenues will first cover all the expenses that we incurred on patenting experiences, prototyping those types of things. After those expenses are recovered, then all future revenues will be split up. My office will take a 15 percent kind of administration fee for ongoing operations. And then after that, the remaining 85 percent is split up; 50 percent goes to the inventor. It is theirs to do privately however they want."

Something about this conversation of percentages started to feel almost dirty. As if the purity of science has somehow been wrecked or tainted from some uncorrupted fiction. My sociological training kicked in, and I immediately felt suspicious of scientific integrity. Obviously Darwin, who wrote and asked his father for money for his trip, was funded to go on the *Beagle*. So his class privilege, in part, protected him (and me), from going on "*The Beagle Voyage*, brought to you by Cadbury." Science has always been commercial, as the work of science historian Steven Shapin attests; there wasn't an innocent time of pure discovery before financialization ruined it.[37] However, as economic historian Philip Mirowski adds, the rise of neoliberalism has elevated the market to being a better processor of information than human beings are; the market, rather than methodological rigor, validates all knowledge claims.[38] USU is a player in multiple markets in which competitors (corporations, venture capitalists, and other universities) put pressure on the Spider Silk Lab by offering more attractive prices and funding options. I'm not suggesting the Spider Silk Lab engages in malfeasance. But the socioeconomic environment that pressures science to be increasingly profitable creates the conditions under which science is conceived and accomplished.

I asked Christian where he thought the goats fit into his role as organizing USU's technology-generated intellectual property. He replied, "So goats are probably our most stable generator of protein. We can consistently produce proteins, we can harvest those proteins, we can purify those proteins, but I'm not sure beyond that."

I asked him to clarify. "How do you scale up the goat herd?"

He replied, "Find enough land and resources like that. Goats have been the most stable line that's been ongoing at our farm for years, back to Wyoming days. It's been a very viable academic option, but I don't know if it is a commercial option." Considerations around production efficiency, as required by capital markets, permeate the narratives of these scientists.

Once the spider silk is successfully innovated to produce a viable object (a jacket, a medical insert, a suture), the goats will need to be replaced by another system—a system that can cheaply mass-produce spider silk protein. Until then, the scientists must rely on the goats to "put grams of protein on the bench," as Justin said, for the continued work of finding viable applications for human use. "Putting grams on the bench" means purifying the milk to capture as much spider silk protein as possible to provide the maximum amount of material for scientific experimentation and innovation.

Systematizing

Perhaps in response to this competitive marketplace, the scientific innovation at the USU spider silk lab, much like the goats, has been systematized, bureaucratized, and monetized. Perhaps as the world becomes increasingly legible through science, the more disenchanted I become. "I don't get it" turns out to be too simple an alibi for my dissatisfaction with interpreting the lab setting only in terms of RNA transcriptions of DNA, or how the mechanical purification of goat milk results in spider silk. Highly specialized applications of science are understood as science itself. In an epistemological irony, as the world becomes increasingly knowable, it is actually harder to make sense of it because scientific knowledges become increasingly specialized. In other words, it is not just the ever-growing scale of the knowable world (big-picture Darwin versus the small, highly repetitive tasks of Grace) but also about how the more we try to understand, the more we realize how little we understand and see how vulnerable we are. Ever-refined scientific innovation is not so much mastery over the world but about our species' lack of power to really control all the variables.

I want to be clear. It isn't Grace's work or that of the spider silk scientists doing these procedures in this lab that bothers me. It is something about the meta-level changes in how I believe we interact with the natural world today versus my Darwinian fantasies. I am embarrassed by what could be described as my Pollyanna wishes (my sentimentality). But contemporary scientific enterprise seems like a quest to write codes and refine techniques—this feels akin to an assembly line. In an attempt to make sellable things, we create solutions to urinary tract infections (Grace's research) or spider silk products. Regardless of the social worth of these solutions, in their creation, there is a process of reduction.

Take the goat, for example. Darwin's understanding of goats (or spiders) would involve seeing and documenting a holistic and immense goaty range of variation: types of hooves, head shapes, eye colors, coat coarseness. These variables are about perceiving what the goat is—the possibilities of goat in all its mutability. In becoming spider goats at the Spider Silk Lab, goats have become a system (among other systems of production, like *E. coli*, silkworms, and alfalfa).[39] As part of the transformation of the spider goats into a system, they are reduced to quantitative measurements. The system must be optimized. The goat is changed both materially and discursively to be more predictable. Humans come to understand the goat differently—I know I do. The goat is now measured in defined categories: its transgenic status (Y/N), its production of silk in grams per liter, its type of protein (MaSp 1 or MaSp 2), the number of kids thrown, the number of female kids thrown. They, like us, become data. But the goats are more than our data. They are also a system to produce a new raw material—spider silk protein. As the scientists modify the "natural" nontransgenic goat, they turn their observations and measurement (their data collection) into an accounting of their own modifications. Is it more accurate to say that the scientists are observing themselves rather than observing the spider goats?

Fordism, ergonomics, Taylorism, outcomes assessments, and program evaluation are human inventions—also systems—that have trickled into our everyday lives. These processes of calculated extraction are not new. In the United States, historical scholarship on these processes are traced back to the violence of early American exploitation during slavery.[40]

We are how many words per minute we can type, how many likes we get, our body mass index, our credit score, our criminal record, our voting habits, our Google scholar ranking, our diagnoses and medication. So it is not surprising that this system, the goat, is becoming a cog in a machine. Karl Marx describes how humans become alienated when they labor under capitalism. No longer laboring for use value (e.g., I make a salad so that I can eat it) but doing so for exchange value (e.g., I make a salad so my boss can sell it), humans become alienated from the process of laboring, the product of their labor, other humans (who are now either bosses or competitors), and, thus, their own humanity, their species being. We become our relation to the bottom line, aligned with ideologies that are not in our best interests.

In particular, we become alienated from our species being, our own humanness, the essence of what it is to be human, which for Marx was our capacity to conceive before producing—to imagine something and then make it happen. I experience this alienation acutely as I age and feel more swallowed up by processes of evaluating everything for capital gain. But can this term, *alienation*, be applied to goats (however mediated through a human lens)? How have goats become alienated from their goatness in becoming a system measured by data points—a system where does and kids don't recognize each other, where mechanical milking nozzles are more familiar than a kid's lips, and where goats live in an open-air pen with no free-range browsing (as goats are browsers more than grazers)? What are the ethical considerations to the practice of creating a new life form that has not existed before?[41] What is our responsibility to that new life form? What is the point of that life form? Should we keep a few spider goats around even after they have been surpassed as cost-effective for making protein? If you bring something into this world, can you really take it out? We have created transgenic goats that are different from goats in the Himalayas, but I resist the temptation to evaluate one group as better than the other. I think this is wrong in a scholarly sense. It's too simple an analysis.

Philosopher Katherine Perlo applies the Marxist concept of class—a category of "any mistreated group of human beings"—to nonhuman animals.[42] Transgenic goats are then a class that can experience alienation. Borrowing from Dutch anthropologist Barbara Noske's reimagining of the four types of Marxist alienation, we can see spider goats as alienated (1) from the product, having many young that are taken away, and milked for exclusive human use; (2) from productive activity, where one bodily skill is forced to specialized—whereas production of grams of protein per liter is the animal's identity; (3) from fellow animals, where animals relations are distorted through confinement and separation, and (4) in total, from their own species-ness.[43] Environmental sociologist Peter Dickens argues from a Marxist perspective that humans use science to modify nature, and through the blurring of science with industry, nature's imperative is to become profitable.[44] Through capitalism, human and animal nature has been transformed. Spiders and goats are fragmented and reconstituted into a new, more profitable creature—the spider goat.

This ultimate quantification of the goat's life is dehumanizing, or "degoatifying." We are no longer companion animals.[45] The goat is a system, a device, an apparatus to augment us. I don't mean to suggest that goats were ever pure; we raised each other up, co-constituted one another in a long history of entanglement. Goats in and of themselves have been yoked for human nutrition, expansion, clothing, and commerce, but now they become newly useful to humans as a machine to get a product not of the goats.[46] Transgenic goats are cyborgs, transformed by human engineering; they are hybrid fusions, "impure creatures" merging two seemingly opposite things: nature and culture, animal and machine, authentic and synthetic.[47]

I hope that my ethnographic work demonstrates how capital is constituted in the entanglement. Capital is not a thing apart but is only imagined to be so (and capital is capital to begin with, because people make it so through material conditions and ideology that reinforces those conditions). Originally I stated that the goats were capital as commodities, laborers, and assets—but this is not specific to spider goats. For spider goats, capital can also be genetic code, silk milk, scientific apparatus (funding structures, networks of scientists, military support, a certain disposition toward invention), and the farming technologies and personnel that keep goats available for this production process.[48] Humans have expanded spider goats' capital potential through the very real microinteractions I observed in the field and the metastructures that sustain it.

From spider webs to goat milk back to spider silk protein and the reconstitution of this protein into different things—these are all acts of translation and manipulation. However, there is no translation or manipulation without purification. In the lab, scientists developed methods to "purify" milk to pull out and isolate the spider silk. Initially this term, *purification*, seemed strange to me. Unlike purification, pasteurization is a familiar process, albeit controversial to some. With pasteurization, milk is said to be made safe for human consumption. Through scientific and cultural work, humans have been convinced that pasteurized milk is more hygienic.[49] Mild heat is used to kill harmful pathogens. Purification, while also a processing of milk, is not to make consumption of the milk safer for humans. Instead, it extracts the protein from the milk, and then the milk is poured down the drain.

To review, spider silk is deemed useful to human applications, in particular, potential military and medical applications, but spiders do not naturally make silk on the scale necessary for human use (or exchange). Because spiders cannot be domesticated or modified to produce the desired quantity, scientists have genetically modified goats to lactate an extra spider silk protein in their milk. To generate the milk, goats must be in a constant cycle of being pregnant, lactating, or being dried off (in preparation to get pregnant again or to be culled). Their milk is taken and, as I described, purified to extract the spider silk protein. At first sight, there is nothing spectacular about this material, but as we will see in the next chapter, promises have been made. From these promises, financial commitments are put into place. Spider silk has a mighty task.

4

Thin Skinned

The Promise of Spider Silk Products

Before COVID-19 closed her school, Greta, my then fifth-grader, proudly bragged about her special job at school. She was the water runner, tasked with immediately running the faucet for fifteen minutes when she arrived at the classroom to flush the pipes of lead. I tried to conceal my horrified eye rolls from her every time she shared this job with one of my friends. In fourth grade, things were so much simpler; she was the pencil sharpener for the entire class. I suppose I shouldn't be troubled by this new job since, like her older sisters, Greta also participates in fairly frequent active-shooter lockdown drills. While I was working on this book, she'd been in three drills. "We go to a corner of the room farthest from the door, crouch down, and stay really quiet," she told me, almost bored with my request for a description, almost as if to say, "C'mon, Mom, you know this." It's all very ordinary, this flushing of our water to avoid toxic heavy metals or preparing and drilling our children for domestic terrorism.

As is the case for other middle-schoolers, the openings and closings of New York City Public Schools over the past fourteen months has taken a toll on her. Considering her newest return to school, I wonder, should we get a face shield in addition to a mask? How much hand sanitizer should she have in her backpack? What other ways will this situation soon become very ordinary, preparing our children to go to manage the toxic metals, potential gunshots, and now aerosol droplets? This world Greta trains for is the same world where transgenic goats have value for their commercial promise. She's certainly becoming attenuated to assessing a variety of risks at an earlier age than I ever was when I had to face such serious dilemmas.

Over the last twenty years, fifteen generations of transgenic spider goats have been bred into existence. Beginning in Ontario, then moving to

Wyoming, and now living in Utah, these goats have been raised by a group of scientists and herdsmen who wish to collect spider silk protein from transgenic goat milk.

Over the years, the project has drawn its primary funding from various branches of the US military, with an eye toward products that defend against bullets and shrapnel that pierce human flesh. Additionally, the National Science Foundation and the National Institutes of Health have also provided funding for projects. Clearly, much of the military and federal funding of spider silk science was based on the starting point of protecting soldiers from explosives and bullets using something "stronger than Kevlar."[1] But Kevlar is not just for soldiers anymore. As the company Active Violence Solutions (AVS), one of several retailers of bulletproof armor, suggests, "The list of bulletproof users is growing." A booming industry, spider silk may replace Kevlar vests and make more than soldiers and police officers impenetrable. AVS offers this mind-boggling list:

Contract security guards
Private security
Executive protection agents
Private investigators
Process servers
Uber, Lyft, and taxi drivers
Bouncers
Bankers
Quick check cashing tellers
Convenience store cashiers
Business owners who make frequent ATM deposits
Landlords and rent collectors
Home invasion protection
Community disaster volunteers
Tow truck drivers
Vehicle repo operators
Pharmacists
Realtors
EMTs and paramedics

Police officers
Community police volunteers
Military recruiters
Firearms instructors
Target shooters
Hunters
School teachers and administrators
DJs
Company executives and human resource managers
Social workers
CPS investigators
Correction officers
Probation officers
Liquor store and late night restaurant staff
Animal control officers
Drug rehab clinic and housing staff

Homeless shelter staff	College students
Amazon and home package delivery providers	K–12 students
	Bull riders and rodeo clowns
Jewelers	Church ushers and clergy
Pawn shop staff	Travelers to foreign countries
Parking enforcement	Customs officers
Restraining order protectees	TSA screeners
State and local park employees	Judges and court personnel
Forest service employees	Attorneys
TV news reporters and print journalists	Elected officials and high profile politicians
Photographers	Celebrities
Witness protection participants	Nurses, doctors, and ER staff
Bounty hunters	Inner city public utility workers
Concert and event staff	Metro operators and bus drivers

Clearly everyone is potentially vulnerable to some form of shrapnel, increasingly so in domestic, civil environments beyond militarized zones.[2] We are no longer secure in public and private spaces, and we're encouraged to invest in some terrorism emergency preparedness kits.[3] Talk about a risk society! I would argue this insecurity has become more and more a norm and that defensive responses have primed the ground for preparedness. As the AVS site also suggests, children's backpacks can be modified with bulletproof panels. (Fear generates both innovation and profit, but what came first?) I imagine Greta as the first kid at her school with a specially fitted backpack constructed to become a pop-up tent of bulletproof material and pulling her friends inside for cover. Perhaps as wearing a shield and mask becomes normal, a bulletproof backpack will be what's coming next for her and for all of us. Maybe spider silk face masks are on the horizon. Disaster capitalism finds opportunities to accessorize wherever it can find even the slightest risk.

Beyond the military applications, scientific researchers hope to innovate applications of the protein into products that can be commercialized in biomedicine. As is common with science, along the way, unintended scientific discoveries expanded the potential of spider silk protein to be useful to humans. Spider silk is also becoming a platform available to for-profit ventures in the human apparel market.

This chapter uses sociological analysis to consider these various spider silk protein products.[4] It primarily explores the ways scientists (both academic and commercial) maintain optimism and generate new applications of spider silk, especially since the products are mostly speculative. Even if a prototype has been constructed, it remains to be seen if products can be approved for use by regulatory agencies, mass-produced, and commercially successful. At this point, the uses of spider silk and the associated innovations are more of a dream than an actuality. Despite the extreme challenges of creating spider silk products, I provide an analysis of these hopeful promises made by corporate, pharmaceutical, and military agents.

Before addressing the spider silk products, both those proposed and those actualized, I interpret the human motivations for modifying these animals and why they have continued to be reproduced through multiple decades. I describe my own attempts to visualize and map the spider goat trajectory, speculating about the larger desire for such a creature within the capitalist neoliberal market. I examine pre-oedipal fears, desires for safety, and yearnings to protect somatic boundaries. This chapter concludes with an analysis of a few products generated from spider silk protein.

Mapping

I've adapted different models of visualizing these overlapping and interconnected parts as a heuristic device. I use these mappings to generate new insights and either to simplify the story or to depict its complexity in new modalities.[5] In so doing, I have created alternative representations of the story. Here are two examples. First is a follow-the-spider-silk approach, linking the various parts of the model in a semichronological order:

Spider (tropical locales) → Spider silk → Funders (US military, governmental, or biopharma) → Scientists → Silking of spiders → Spider silk protein → Spider DNA → FDA → USDA → University (Wyoming, Utah) → Transgenic goat → Buck → Kid → Milk → Milking machine → Herdsmen → Veterinarians → Lab workers → Purification machine → Spider silk protein pellet → Experimentation and innovation → Commercial partners → "Products" → Media ↔ Feedback loop ↔ Public → Spiders, goats, and humans

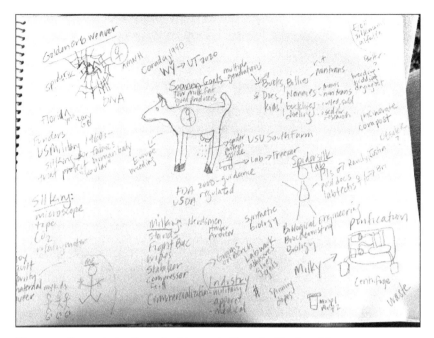

Figure 4.1. Conceptual collage of spider goats. Photo by Lisa Jean Moore.

Another option is a conceptual collage, as shown in the accompanying figure. Dozens of times, I've sketched out new maps while reviewing discarded ones, yearning for a comprehensive diagram of the trajectories, processes, and human and nonhuman actors, but these elements are not static. A one-size-fits-all approach is impossible. With the help of the maps, I have been better able to focus on the *how*, as the previous chapters have outlined: how the transgenic goats have come into being. I've broken down how this multispecies endeavor arrives at a product. The making of the transgenic goat transforms not only the goat but also the spider and the human (among other objects). The goat and spider become kin. The spider becomes the source, the pristine expression of the silk, the object we are trying to get back to through all this modification. The human becomes a transgenic tinkerer, further enmeshed with a multispecies network, modifying the phylogenetic tree.

Moving on from the how, I found the *why* trickier. The simple answer is that human ingenuity is fueled by desire. However, persistently skeptical sociologists propose that capitalist incentives shift human consciousness so much that we can never really be sure what is an authentic desire (shown as a feedback loop in the previously described model). Rather, the market has been so fully integrated into our consciousness that our desires are algorithms of sophisticated data mining of our demographic—if you liked *Twilight*, then you'll love *Fifty Shades of Grey*.[6] Humans have felt that spider silk is necessary for reasons beyond its utility. It has physical and magical qualities, not all of them real or material, to which we become attached or stuck. Their use value is entangled, to pun further, with our arachno-imaginaries.

Flexible Fibers

There is clearly a material promise to spider silk. On the surface, the promise is about a new strong, flexible, affordable fabric or adhesive that protects the human body from harm. Just as spider goats are designed as machines, affordances are built into spider silk itself. Our modern risk society disciplines human actors to anticipate both risks and solutions, and sometimes not even in that order. Spider silk is a product born out of militarized potential to mitigate risk. But transgenically produced spider silk is also in many ways a product in search of an application; it's as if scientists must anticipate an existing problem or a potential risk for which spider silk would be the best solution.

The creation of transgenic goats and the expression of spider silk protein is brought into being by heterogeneous actors offering optimistic assurances of what spider silk will do. Justin Jones shared with me some of the potential uses his team considered for spider silk proteins:

> [An] abbreviated list of items (nonfiber) we have looked at for goat-derived spider silk proteins:
> - Adhesives
> - Catheter coatings that reduce thrombotic occlusion, fibrin deposition and infections associated with IV or urinary catheter
> - Medical device coatings
> - Drug delivery nanoparticles
> - Drug eluting/wound repair hydrogels

He continued, "None of these items are products yet in the sense of being sold. The only spider silk available on the market in a product is being used in cosmetics from Bolt Threads and AMSilk." These two for-profit companies make products that use biomaterial fiber; I return to one such company at the end of this chapter.[7]

Furthermore, at some point during the three years that I was conducting research for this book, these and other items were suggested as applications of spider silk protein inclusive of fibers (most commonly produced by silkworms):

- Bulletproof clothing
- Parachute cords and parachutes
- Arresting gear ropes to catch a plane tailhook on aircraft carriers
- Active apparel, running shoes, jackets
- Seat belts
- Mesh covering of insertable devices in human bodies—catheter coverings, vaginal mesh, hernia repair mesh, mechanical insulin pump coverings
- Biodegradable bottles
- Adhesives
- Sutures, bandages, surgical thread
- Artificial tendons or ligaments
- Applications in robotics
- Skin-care products

While none of these items is actually in production today, there have been a few prototypes, and I've considered what these promised items have in common. A promise is a future not guaranteed, so it's a dream. They are all things that protect humans. Spider silk is capable of enveloping the human body in a powerful security blanket, a supposedly stronger-than-steel membrane over our fragile, flimsy skin. It can also be a waterproof antibacterial coating adhering to devices inserted inside our bodies and warding off nasty contagion—to consider it psychoanalytically, we must protect the pure from contamination. It keeps us safe from harm in dangerous situations. In the risk society, its capacities are only limited by the innovators' vision.

So we hear proposals that spider silk promises us better medical treatment in the form of adhesives that perform better than others on the market. During an interview I conducted in the Spider Silk Lab, graduate student Danielle Gasinby discussed the application of spider silk adhesive for bedsores and burns:

> We did bandaging using these spider silk proteins as a spider silk skin adhesive. We basically made a second skin layer using the electro spinner. And it basically draped over [the test object], and we would secure it with a spider silk adhesive. And we were using the goat milk spider silk because it is the purest. The properties of our second skin were very close to the properties of actual skin. That was really cool. And we found that some of these applications for the medical field were very good. The adhesives of spider silk are very elastic. So as an adhesive, it will stretch before it breaks. So therefore, it is going to stay on longer.

A simple trajectory from Danielle's remarks could be to desire a second skin, a layer of protection, and to be cared for in the most efficient way possible. Spider silk promises us medical treatment. More than that, it promises us remedy, something that relieves or cures a body disorder. Under capitalism, the value of spider silk is its economic utility. We devised a better method for making spider silk, we feel ownership over our new and improved spider silk, and thus we have a right to use it for an economic end. But the promises are not all about potential utility.

Stepping back to consider this unrealized, yet long-standing, desire for strong, protective coating unmasks us as self-conscious of our status as vulnerable, puny animals. For centuries, we have exploited the capacities of other animals in our quest for security. Beyond animal testing to develop interventions, drug regimens, or treatment procedures, animal bodies are harvested for raw materials to protect human beings. Horseshoe crab blood is used to test our insertable and injectable medical devices for endotoxins and is essential in vaccine production, including a vaccine to end the COVID-19 pandemic. Mammals are skinned to produce hides to clothe our exposed skin. In the case of spider silk protein, there is the hope we can innovate materials that will provide safety from shrapnel, from bacteria, from bullets, from infection, from deterioration.

As established in chapter 1, and as a plot twist, one irony is humans' fear of spiders is turned on its head with spiders as our saviors. Beyond arachnophobia, I am drawn to other theories that might offer explanatory angles of why humans modified goats. Are we the only animal that uses other animals as machines to make products to protect us from ourselves, our own human-made threats to our well-being? Clearly, we are part of the machines, as we're made of animals. They are in our blood, and we mechanize our actions for them as workers, DNA carriers, microbial homes.

Psychoanalytic theory appeals to me here as a modality of thinking through human motivations beyond commercial opportunism. I am compelled by how the lab community exudes an earnestness in the desire, a hope, to create useful objects to protect bodies. What drives this hope? Could an interpretation of primal fears enable an enriched understanding? For me, it is no coincidence that psychoanalysis could be on my mind as I write this book and struggle with the challenges of motherhood (specifically, raising three daughters) as I myself go through menopause, my parents' illnesses, aging. I search for ways to make sense of my life as I try to make sense of my field site. How do I dig deeper to figure out new ways to comprehend or represent the world? Sigmund Freud is often brought to mind as the originator of psychoanalysis; the field proposes that humans have very early primitive anxieties that unconsciously drive them to enact particular patterns of behavior. Psychoanalysis encompasses methods and theories that guide how to reveal what might reside in the unconscious.[8]

Donald Winnicott, a British pediatrician and psychoanalyst, has been a favorite of mine for the past twenty years or so, primarily because of his theorizing about the "good enough mother."[9] Writing from the 1950s through 1972, he suggested that mothers who fail their children (in reasonable ways) are actually enabling their children to tolerate the challenges of everyday life. To be a perfect mother is detrimental to the mother and the child in that the child develops unrealistic expectations about the world and doesn't develop skills necessary to tolerate disappointment and frustration. (And the mother winds up exhausted and possibly resentful and emotionally drained.) For obvious reasons, the allure of Winnicott provides a much-needed give-yourself-a-break reminder when maternal self-recrimination kicks into high gear. He is also known for coining the term *transitional object*, that smelly blanket or stuffed animal belonging to young children and possessing some fantastic quality, often enabling the child to fall asleep (as any parent who has left "Bunny" on the airplane will attest).

In "Fear of Breakdown," Winnicott's last essay (published posthumously), he explains that fear of breakdown is a "universal phenomenon" wherein the unit itself experiences a breakdown in the mother-infant bond.[10] Winnicott proposes five primitive agonies that an infant experiences when they are confronted with a traumatic experience, an "early annihilation experience," or the lack of a good-enough mother-child bond.[11] For example, one of these agonies is a fear of returning to an unintegrated state, which I liken to being fearful of being shattered, ruptured, or leaky—the feeling of needing to shore up yourself, your identity, your body boundaries. Psychoanalysts sometimes refer to the fear of breakdown as *disintegration anxiety* and consider it the greatest human fear.[12]

One interesting fact about the fear of breakdown is that even though it might have been experienced in the past, it stays with us in our unconscious. Winnicott writes, "There are moments, according to my experience, when a patient needs to be told that the breakdown, a fear of which destroys his or her life, has already been. It is a fact that is carried around hidden in the unconscious."[13] So we carry around with us "unassimilated traumatic experiences."[14] Human beings are confronting the fear of breakdown. We endlessly work out these fears in everyday life—no matter how healthy we may appear to be.

Perhaps we humans work out these fears expressed in primitive agonies deeply entangled within our psyches by making things to protect ourselves. Humans are afraid of the breakdown of their own organic coverings, afraid of mutilation, scared of body-boundary breaches, and terrified of foreign objects entering their bodies. I still recall watching my just-walking toddler see her own blood from a scrape and howling with fear, "All of me ... coming out!" A need for an immediate Band-Aid to patch up the leak was imperative and instantly soothing. The terror of annihilation is a primal human fear, but people with good-enough caretakers don't experience the terror in the way others do (whose boundaries have been breached). On the other hand, human life is inherently fragile and our psyches work hard to keep that reality at bay most of the time. It is remarkable that the fantasy/wish of spider silk propels this ongoing research, despite the easy availability of other market-ready objects.

This primal fear that our bodies will be hurt means we go to great lengths to protect our bodies—even if the harm is self-inflicted from other members of our species, as in a school shooting. We reengineer, genetically alter, harvest, extract, and otherwise tinker with other species as machines to produce the raw materials to keep us safe. (We are also preoccupied with harming our own bodies or at least experimenting with the pleasures and dangers of self-inflicted wounds.) This cycle of safety and peril is then a boon to shrewd venture capitalists honing in on fear (and sometimes manufacturing it) to conduct brisk trade in disaster and danger capitalism.

I am not suggesting that all spider silk protein innovators have had early annihilation events and traumas that led them to dwell for more than fifty years (since at least the army's 1968 report) on the quest for a protective film. But I am suggesting that as a species, human beings toil in extraordinary ways, perhaps driven by underlying primitive agonies that motivate innovation: our desire to be safe or to ward off harm, our hope at discovering the ultimate protection. Because we as a species have collectively experienced trauma, we go to lengths to produce things that make us safe. What is the collective trauma or emotional catastrophe that has produced a need for spider silk? Perhaps our desire to be safe, to not be breached, or to remain whole and intact stemmed from a fear, a primitive agony that has generated a desire for protection. This fear has spawned a great deal of bench science, while no bulletproof vest made of spider silk has been created. In the sections that follow, I explore what has emerged instead of the protective gear.

Dental Adhesives and Catheters

The goats were invented for a specific purpose, to synthesize and produce spider silk protein for manufacturing military-grade fabrics. But over time, that purpose evolved. That's not how research and development or, for that matter, capitalism works. Research is wobbly, sometimes accidental, and sometimes purposeful; it produces indirect failures that generate new twists and inventions. If the research were all utilitarian, then it would have ended long ago because the desired results just weren't achieved. But still the work slogs on, and the scientists try to get more money to figure out how to make spider silk affordable and useful.

Since the beginning of the new century, other possible applications of spider silk have become apparent through scientific discovery and innovation. In the Spider Silk Lab, there is a sense of urgency about creating an affordable and marketable product that can be mass-produced. As discussed, there are four systems used to create spider silk protein—goats, *E. coli*, alfalfa, and silkworms.[15] The lab is divided into teams that are simultaneously extracting spider silk and innovating products. For the purposes of my project, I am focusing on the spiders and the goats, partly because the goats were the first to be modified as a production system. As Justin put it, "I think that everything we have done from an application or research standpoint for spider silk protein has stood on the shoulders of these goats." The goats (and the spiders) were foundational for all spider silk protein research.

As discussed in previous chapters, the goats express two proteins of major ampullate spider silk, MaSp1 and MaSp2. I asked Bri to explain the difference to me.

"So the MaSp1 is traditionally used for very stretchy materials," she said, "while the MaSp2 is more for the rigid materials." She continued telling a story about the serendipity of science. "MaSp 2 is also very good for making adhesives or polymers. We actually found out it makes truly good glue. We found out about our glue through mistakes. It was a happy mistake, so that's lovely. We kept gluing syringes together, and then we realized it had this great property." From these moments of discovery in the lab, prototypes of objects are made to attract commercial partners. In the case of the Spider Silk Lab, dental adhesives and catheter coating are two such biomedical products that have faced difficulties in breaking into the marketplace.[16]

During my interview with Christian Iverson, the previously mentioned director of Technology Transfer Services at USU, I asked him, "What do you see as some applications of the adhesives, polymers, or gels?"

He replied, "Dental adhesives for periodontal disease. We've had a lot of grant funding, and that's a pretty interesting, unique one. Now we're kind of looking for more of a dental partner. Preliminary data shows it's pretty impressive and it could be a pretty unique and beneficial treatment. We've got some commercial partners lined up that would be interested in it once we have an initial data set. We have a dental school in Salt Lake that's interested in collaborating with us. It's just finding the right pathway."

In another interview, Justin told me that while the lab can make a device for small-molecule drug delivery, making enough of the product is a challenge. "We aren't going to go through FDA approval for something when the fragility of our system means we can't scale up."[17]

A web page from the Technology Transfer Services group says that For this application, spider silk gel could be used "as a vehicle for periodontitis medications and are applied to infected oral areas in the form of a gel or solid structure. The properties of spider silk allow the solution to stick to the infected areas for long periods of time and slowly release the treatment, making it much more effective."[18] It's hypoallergenic and therefore biocompatible with human bodies and safe for human use.

Christian was careful not to exaggerate the spider silk claims. "I am not thinking of any large-scale construction projects," he said. "Everybody wants the low-volume, high-value product. So, that's why we're looking at the kind of medical fields for those. We visited with some companies probably two or three years ago. Another thing is suture replacements. So those types of things, instead of using sutures, they would be adhering with silk. And then, you know, we've seen some preliminary results on cell growth based upon growing on the surfaces."

I pushed, suggesting some type of larger-scale spider silk bionic intervention. "Could we have knee replacement or hip replacement made of silk?"

Christian put his hands up in a "Whoa, Nelly" gesture and said, "Yeah, I don't know if I would see it made from spider silk. But what I would see is a coating. There's still a lot of testing we need, to go into something like this, but preliminary information and data shows that it's non-immunogenic, and this is useful for a catheter." He gestures stretching out a thin tube. "So we are here with the catheter-coating project that showed a reduction in the amount of biofilm formation. And so, is [spider silk protein] preventing [microbial growth]? Or is it antimicrobial or is it preventing surface adhesion? We need more research, but it's promising."

I felt a bit as if I were getting a sales pitch.

Although the research on catheters has energized the Spider Silk Lab, Randy expressed his frustration at trying to move the product into practical use. I noticed a real difference in his mood between my first visit to Utah, in 2017, and the subsequent one in 2019. On the second visit, he seemed more cautious and less gung ho than earlier. I mentioned this observation to Randy and asked, "Are you feeling differently about your work?" I was nervous to ask this, as it veered into a different type of interview question than the purely technical.

But he replied with his usual candor. "I guess I'm less optimistic than I was a few years ago, and I think part of that is because we know there are applications that make good sense for this and we just simply are not able to get that done. So coatings for catheters make perfect sense, same as meshes for hernia repair."

I nodded sympathetically, and I silently reflected on how Randy had turned from energized to disappointed over the course of my fieldwork.

He continued, "We had a group that very clearly demonstrated the stuff we're making is better than what's out there right now. But the industry is loath to touch meshes, because of all the issues they've had with vaginal meshes—lawsuits and all that kind of stuff.[19] So, they'll put in something . . . plastic that for many people is completely too thick. It's too inflexible. It's got all kinds of issues with it, but they know the FDA approved it. So they can't get sued for something they have approved." He shakes his head, and shrugs as if to say, "Go figure."

Here Randy was referring to his frustration (and that of other innovators) with the FDA's approval process for new medical devices. Biopharmaceutical companies often don't want to invest in an expensive and intensive approval process for new technology and hardware if they have something already available that works, even if it is inferior to what is newly available.

He explained further, "Two of the biggest manufacturers of catheters are in Salt Lake. And we've talked to both of them. And both of them are, like, you know, 'We can't raise the price of our catheters by one dollar.' Or 'The hospitals will not buy it.' And . . . even if we save them thirty thousand dollars per patient who gets an infection, because for older patients, they have to pay for Medicare, but who will pay for it if you pick up the infection in the hospital? So it's cost-effective at even ten dollars more per catheter; that's three thousand patients that you're not going to be paying for if they were to get an infection. It is preventative and cost-efficient."

But capitalism isn't a system built for long-term thinking or future savings, I wanted to tell him.

"Like, how does that not add up logically?" Randy asked. He paused and placed his hands on his desk for emphasis. "We've been down there and seeing their plant process. We could very easily incorporate what we need to do into that process, and they admitted that they just said, you know, 'Our people can't sell a catheter for a dollar more than somebody else's.' Even if I'd said, you know, 'We'll cut down your infection rate by 70 percent,' [it] doesn't matter. They don't care. That's not something that they're going to use for sales."

Justin had a different perspective on why catheter coatings were not viable at that time. He described how the lab conducted a survey of catheter manufacturers. It became abundantly apparent that the lab would never find a partner until some facility could manufacture a catheter that would not shed any of the spider silk protein.

I could see Randy and Justin's predicament. It is hard to get medical industries to take economic or biomedical risks (but not so hard, apparently, to take risks with women's bodies). Catheter coatings require a fiscal investment to get approved, and meshes are a hard sell at this particular elevated movement. I empathized with both of them. "So in order to continue to do the science, you have to spend a lot of time writing grants?" I asked Justin.

He nodded vigorously. "Yeah, there are venues that would be interested, but it's challenging because people are threatened by trying to do something new. It's going to probably be a company out of the US, because here they just have no interest in the long term."

I interrupted him. "So some other country would be interested?" I asked.

He screwed up his face. "Not exactly," he said. "They'd want us to figure it out. So that, you know, they've got a six- to eighteen-month period to move it along to get it to market. And so they want us to cover all the costs till then. And we're probably geographically in the wrong place, if we were in California." Justin was alluding to the prospect of venture capital and Utah's limitations in this regard, considering its distance from Bay Area capital.

"I mean, you look at both Bolt Threads and you're sitting there going, 'How in the hell do they raise $210 million dollars?' And they're selling purses. They sold ties for $350 made out of spider silk. They had no mechanical properties whatsoever. They're selling watch bands. They're putting it in cosmetics. You know, this has nothing to do with spiders. So it's all just marketing."

This notion of the for-profit sector using spider silk for a variety of nonmedical, nonmilitary products—that is, more consumer goods—indicates the pizzazz of having a new angle to establish buzz in the marketplace. And while I'm ashamed to admit it, I also think spider silk jackets and face creams are sexier than catheter coatings.

The Material You Didn't Know You Needed

The apparel market, although not specifically using goats for their spider silk protein production, has an important place in the landscape of spider silk products promised to entrepreneurs. A short list of some of the companies incorporating spider silk into apparel includes North Face, Adidas, Polartec, Patagonia, and Omega (for watch bands). These companies' products have mostly been prototypes for trade shows or outdoor magazine stories.

The Utah team at the Spider Silk Lab is well aware of the potential commercial uses of spider silk, and as Randy suggested earlier in this chapter, the incorporation of spider silk into consumer products sometimes feels gimmicky. Christian Iverson clearly felt that spider silk was being used "more as a concept rather than any wearable usable outdoor product." Transgenic silkworms are being used to create a fiber that is not 100 percent spider silk but "a drastic improvement over a current silk." Christian also said that "private companies are making some pretty quick advances and probably passing some of us up." He was referring to Bolt Threads. "They've raised between $200 and $250 million or more. It's been private money, venture capital."

There is a desire to change our reliance on certain fabrics. In her book on synthetic biology, Sophia Roosth describes how much of the twentieth century was dominated by an agenda to convince consumers that synthetic plastic fabrics were better than natural fibers.[20] Much of human clothing is made from acrylic, polyester, nylon, and spandex, the production of which is linked to global warming. Ironically, at the beginning of the twenty-first century, our consciousness has been raised to the dangers of petrochemicals. We are shifting back to natural fibers (flax, wool, hemp, cotton), which might well be more sustainable, and potentially vegan materials through innovations in synthetic biology.

Justin concurs with this shift in our beliefs about fabrics. He thinks the Spider Silk Lab has provided the scientific knowledge to create sustainable biomaterials.[21] He told me, "Spider silk, from my perspective, which might not be widely shared in the community, is that the spider silk research might be more important than the spider silk itself." He continued, "Right now we have a huge plastic problem, and we need to move toward biomaterials, but how are you going to produce them? We have provided a great deal of research on how you produce difficult proteins en masse. That might be our longer-lasting contributions."

Although spiders generate waste (as Cheryl Hayashi said, "spiders themselves produce the typical waste associated with being living organisms, carbon dioxide as well as liquid and solid waste [poop]"), spider silk itself can be characterized as recyclable. Cheryl explained that spider silk is not biodegradable "anecdotally, spider silk appears to be resistant to biodegradation (which I am narrowly defining as breakdown by naturally occurring microbes in spider habitats). In fact, there was a recent paper on spider silk inhibiting microbe growth due to low nitrogen content."[22] However, spider silk is recyclable, she said. "Silk is proteinaceous, so [it] can be broken down by enzymes (think digestion), allowing the constituent amino acids to be reused." Consequently, spiders can eat their own webs and use the protein to make more webs. Humans can also recycle threads in the manufacturing of fabrics.

One such commercial company staking its claim in the biomaterials market place is Bolt Threads, launched in 2009. In September 2019, I spoke with David Breslauer, the cofounder and chief scientific officer of Bolt Threads. David earned his bachelor of science degree in bioengineering from UC San Diego in 2005 and PhD in bioengineering from UC Berkeley in 2010. His graduate work focused on the material science of spider silk fiber production. At one point, Cheryl Hayashi taught him how to silk spiders. At Bolt, he has built the R&D teams delivering Microsilk and Mylo for apparel and b-silk proteins for beauty. I explained to David my keen interest in the role of the goats as the initial system to express the spider silk protein. He quickly responded, "Randy's research was absolutely the fundamental work on spider silk. And because of Randy's goats and the fact that they could make a gram of protein at the time, that research helped us understand silk in ways we never could have before."

David spoke with enthusiasm. "If you dig into Nexia business, I found ten years ago a business proposal about goats to be used for protein pharmaceuticals. All their ideas were right, only goats were the wrong core technology. The goat has to be replaced with microbes."

I jotted down a note about the wrong core technology and circle the word *goat* on the page.

He continued, "Spiders, that's an interesting thread. I am obsessed with spider silk in its own way, and the more I did, we really started digging into the consumer apparel market. We can engineer protein from the spiders, and it became really apparent that all the environmental challenges of the textile world were horrifying in their own way. What you come to realize is the number of places where biomaterials can have an impact, so our mission kind of grew. We were part of making biomaterials in a push not to be petroleum based—make them amenable to modified chemistry that [is] environmentally friendly." Bolt Threads uses yeast as a system to create its spider silk proteins. In this way, the company's materials explain its innovation of Microsilk, a fabric inspired by spider silk and manufactured using bioengineering to put silk genes into yeast. The silk protein is then captured through fermentation, purified, and spun into fibers.

Online, David's bio reads with a flair and a nod toward marketing and branding: "Initially based on the graduate work of myself and my other two co-founders studying the material science of spider silk, Bolt Threads is working towards our shared vision of harnessing the astounding materials found in nature to build products that combine innovation, performance, and sustainability."[23] He has a different take on the stronger-than-Kevlar motto that has been synonymous with spider silk for generations. "When everyone talks about spider silk," he told me, "the myth that never died is that the spider silk going to make the future of bulletproof vests for soldiers. There's nothing fundamentally wrong about that idea, but there are a lot of reasons why that won't work, and no one's done it. But it gets funding for science, and everyone has heard of it."

The funding for science is based on "national security priorities," David said, but he explained how his team decided to go in a different direction. "Bolt Threads started to make scalable spider silk to do it right in an effective way to make economically scalable amounts. We quickly realized the difficulties in silkworm unpredictability and turned toward molecular biology and then first entered the market with more industrial-produced silkworm silk. We used a spider silk protein polymer in a microbe, taking the same sequence of a spider [silk protein] without the full length [of the protein] to make a silkworm-quality silk fiber minus the biological variability. We think we can make a protein fiber for apparel that replaces silkworm silk." In 2017, Bolt Threads and fashion designer Stella McCartney teamed up to produce the prototype of a shift dress made out of spider silk for an exhibition on clothing for the Museum of Modern Art in New York City.[24]

During our conversation, David shared with me how the partnership with the fashion designer has been good public relations. Through a perceptive and timely understanding of the shift toward biomaterials as well as a rising cultural consciousness around animal rights, Bolt Threads, which did not start off as ethically committed to eliminating exploitation and cruelty to animals, is in fact accidentally vegan. Conventional silk is not vegan, because silkworms are typically killed in the process of collecting their cocoons and unraveling the silk onto spools. He jumped on an opportunity to capitalize on the Stella McCartney brand, he said, because she has "singlehandedly convinced the fashion industry that tech could cool." He continued, "Stella McCartney said it best: 'I don't really care about worms, but if I don't have to kill them, I don't want to.'" McCartney is the daughter of Wings band member and dedicated vegetarian and cookbook author Linda McCartney, who was a well-known animal rights activist. It remains to be seen if any spider silk dresses will be available at retail stores, but in 2019, Adidas and McCartney teamed up to make a biofabric tennis dress using orb-weaver-derived silks in the fabric but synthesized through yeast fermentation.

Back at the Spider Silk Lab, Justin seemed skeptical of Bolt Threads' ability to spin fibers to satisfy a larger apparel industry. "Even if they can do it, the high-end clothing application is a very limited market. This might be why they are moving to face creams."

Launched in 2019, a skin-care line (formerly Eighteen B and currently Beebe Labs) is the Bolt Threads cosmetic spin-off company with proprietary b-silk added to the products.[25] The company describes the process as cellular agriculture, whereby acellular animal products are made without animals through the use of microbes like yeast or bacteria—silk production for a "clean beauty market."[26] According to David, "the spider silk proteins hold on to moisture well and fill in the face with microscopic films that are good at keeping a fine layer of moisture on your skin." However, David is aware of the overwhelming negative associations with spiders. "So we don't advertise it as spider silk on the face. You will find this generally with lab scientists; they don't want to ignore the spider silk and they are so enthusiastic about spiders they want to spread the word. So we have an internal conflict where the scientists are OK with saying it's spider silk on your face, but there is a reason they are not in charge of marketing." Bolt Threads and its spin-off company, Eighteen B, are not the first to use silk in cosmetics, but they claim that since their silk protein is not broken down into smaller components, it is a better product.[27] Significantly, this is the first cosmetic to mimic the properties of spider silk. "Ours has functionality, and it happens to be vegan," David explained.

Eighteen B was a direct-to-consumer company that promoted itself through its website, eighteenb.com. The company has been rebranded as the Beebe Lab. The current website is, and the former website was, aesthetically slick, with many clean and minimalist laboratory images, a microscope, a bespectacled female scientist in a lab coat, a slide of magnified molecules. No DIY-tinkered contraptions here, but a sleek scientifically curated tour of the company's facilities designed by an image consultant. Eighteen B originally sold three skin-care products ranging from $75 to $105, positioning it within the luxury skin-care market. The products claim to "harness the power of silk." Like the stronger-than-Kevlar motto used to promote the product's ability to protect skin from shrapnel, b-silk protein is said to create a "protective barrier" where "better skin is possible through better science" and to "reinforce its [the skin's] extracellular matrix."

This is a "smart solutions" company where wrinkles can be firmed up while securitizing skin. And the market for this skin cream, ironically, is someone who looks a lot like me. An older professional woman

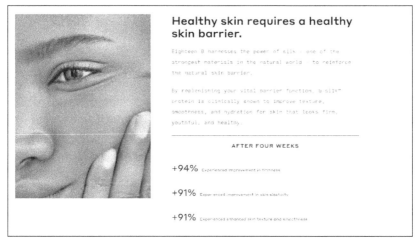

Figure 4.2. Screen shot of Eighteen B promotional materials.

susceptible to green marketing, seduced by claims of natural, organic, cruelty-free products that reinforce health. Clearly not a primitive agony but definitely a continuously stoked cultural anxiety, aging has generated an opportunity to innovate new products for spider silk. Maybe wrinkly skin that signifies aging and a fear of disintegration or withering away (both literally and symbolically) is an advanced or future agony that skin cream can prevent.

From Substance to Style

Throughout the project, I've returned again and again to the question of both why and how goats have been modified and what this modification means. As an analyst, I've immersed myself into different spaces with a give-me-all-you've-got attitude. I take all these different pieces—spider legs tickling my forearm, math equations handwritten by my daughter, plastic tubes shooting out modified milk, baby goats' incessant cries—and put them in contact. I dream about the goats and the spiders and my laptop in weird configurations, and I wake up with some fleeting clarity that fades as I sit down to write. The tactile sensations of my fieldwork are hard to express in text. Methodologically, to keep things organized and productive, which has always been my personal morality tale, I use a variety of strategies to make sense of the emergence of these transgenic goats that synthesize and lactate spider silk protein.[28] I'm also confronted with how this fieldwork has now spilled out into my life—I am trying to do feminist science studies as a menopausal fifty-four-year-old woman, a mom of three young women, a person frantically scrambling to make sense of the world as it dies, melts, sinks, and burns. I worked on this book while the nation continuously seemed on the brink of additional endless wars as a former reality-television personality turned president tweeted hawkish threats. And we now face a zoonosis that has radically transformed all life for two years. My fear is fertile; it gives birth to my anxious thinking, writing, and making. But my fear is also soothed by consuming—face creams, apparel, and medicine.

My reading of social theory, specifically Lauren Berlant's *Cruel Optimism* and Anna Tsing's *Mushroom at the End of the World*, helps me deconstruct the promises of spider silk and the ecological messes we must bear witness to in these precarious times.[29] As Berlant writes, cruel optimism "exists when something you desire is actually an obstacle to your flourishing."[30] Is it possible that spider silk promises us something but will eventually and inevitably let us down? How does its promise end up bad for us? Lurking always in this project is the sinking feeling that this massive innovation of spider silk and goats through the past twenty-plus years will not achieve the types of solutions heralded. Stopping bullets or shrapnel just motivates the creation of better bullets and stronger shrapnel. We can't flourish as a clever species of synthetic biological innovation if wanting spider silk just ups the ante to innovate something to outsmart spider silk.

In a nostalgic version of the spider goat story, necessity is the mother of invention. We needed something, so we made it. And that human ingenuity might have been partly fueled by a deep need for protection. Persistently skeptical sociologists (myself included), however, propose that capitalist incentives shift human consciousness so much that we can never really be sure what is an authentic need (a feedback loop, as indicated above). Which came first, the instinct to protect or the production of fear of being unprotected? Also it's clear that all our wants have been turned into needs.[31] The market is the mother of invention. Wants (or market suggestions) are hegemonically transmitted as needs that define the contours of the good life.

While I am well versed in this type of critical thinking that suspiciously judges all transactions as market driven, I'm interested in a different tack for discerning "why." In other words, I am not trying to skirt a Marxist critique and despite the obligatory 'everything is driven by profit' moral to the story, I do see how human workers (lab techs, herdsmen, scientists, me) make their work meaningful, even if it is driven by the pursuit of capital. Spider silk was deemed useful because it has properties beyond what is human-engineered; it is both strong and flexible and potentially could be used to protect thin-skinned human bodies. But it was needed in large quantities, quantities beyond what can be harvested in the wild. Goats are transgenically modified as the original system, the fount of spider silk protein. Hours, days, months, years of tinkering to create applications that save us, protect us, envelop us. But a large-enough quantity of spider silk protein is too expensive to make—we can only make a very little bit. Small amounts of spider silk are the mother of invention.

Conclusion

Knowing You're a Goat

While spending time with the goats in Utah, I had fleeting fantasies of being that younger, more radical version of myself and setting the goats free. I'd help them, and they would roam like the goats I saw in the Himalayas. I'd celebrate with them being out and about in the expansive landscape all around us. I'd serenade them "The hills are alive" style à la Julie Andrews in *The Sound of Music*. In my fantasy, the goats kick and jump for joy, butting my legs to thank me. Following a Donna Haraway sensibility, I would try to make kin with them.[1]

But spending time with the goats, I realized that even if I opened the pens to free them, they would likely run up to the milking stand and wait to be fed and milked. My savoir status is complicated by their actual behavior—do the goats have false consciousness? Do I? How would I know in either case? Throughout this project, I wonder again and again if it is even possible to actually know you have false consciousness.

Given the rigid FDA and USDA regulations that manage transgenic species, I realize that these goats, their predecessors, and their progeny never have been and never will be untethered. Randy and I chatted about the human fears of transgenic goats breeding with other goats and "taking over the species" and how these fears drive regulations to confine the transgenic goats so they don't overrun goat breeding. People are afraid of transgenics (in the form of the goats) run amok. But Randy believes these are unfounded fears. He explained, "There is no reason to expect that if they were to get out, that they won't be eaten by coyotes, which is the highest probability because our goats have no fear of anything. They are all completely acclimated to humans. Every goat wants to chew on you. They want to be petted; they want to suck on your fingers. So they have no concept that they are really goats." They wouldn't fare well in the wilderness game of survival of the fittest.

I've thought about his comment "They have no concept that they are really goats" many times while I was writing this book. Obviously, this is an anthropogenic projection of what humans, in this case, Randy, believe real goats are. The comment suggests that goats have a pure ontology that domestication, including transgenic breeding, distorts. This line of thinking brings to mind philosopher Thomas Nagel's 1974 famous metaphysical essay "What Is It Like to Be a Bat?"[2] In this essay, Nagel uses the bat to explore the essence of experience and to argue that everything has its own sense of what it is to be itself. To counter reductionist claims, he argues that perception exists and happens in other life forms, even potentially even outside human imagination.[3] Bats use echolocation to perceive their environments, and humans lack this capacity. We, as humans, have no idea what it is like to be a bat, because we do not have the same type of perception that bats have. Instead, we use metaphor to imagine what a bat's experience is. Some truths of what it is like to be a bat can't be learned or experienced by humans. But with the proliferation of COVID-19 theories positing bats as reservoirs of zoonotic disease, there has certainly been an uptick in humans imagining bats' motivation.[4]

Building on Nagel's essay, I ask, What is it like to be a goat? And I must admit that I am severely limited in my ability to know the answer. Even if they experience fear or love or loss, they don't do so the same way that we experience these things.

Despite the presumption that a baseline consciousness of goatness persists, domestication, including transgenic manipulation, changes that goatness, creating new qualitative subjective experiences. But when did each intervention change goats to a new species or subspecies? Is it more accurate to say that goats have an experience of their particular type of goatness? That transgenic goats are a type of goat unto themselves? My thinking bleeds between specific ontological questions I've been raising and the larger metaphysical ones in which they are nested. Does goatness have an experience of life, of *isness*? But does that isness tie into anything greater, nonphysical? In this regard, I both feel anxious about being a competent researcher and manage a metaphysical anxiety that doesn't belong to the goats but belongs to me and my desire to help them. Am I trying to save goat souls?

The nexus of the reproductive and the synthetic disrupts the idealization of scientific knowledge production as rational, clear, tidy, and objective in a reality that is messy, imprecise, unpredictable, and affective. Overlapping threads bind goats, spiders, scientists, lab objects, veterinary science, state actors, academic research centers, entrepreneurs, war machines, fashion designers, and cosmetics in the pursuit of spider silk protein.[5] The material-semiotics of the spider goat (as Karen Barad points out) are concatenous and recursive. That is, the narratives and the material are the making and unmaking, the already-dense and interlocking web of material, narratives, and meanings, starting with the chimera of Greek mythology and weaving through children's books and nursery rhymes, generating lush opportunities for human imagination when we consider the transgenically connected animals: goats and spiders. Strands of genealogy and phylogeny also bind microreproductive communities, including goats and spiders and humans, and specifically me and my offspring. Heterogeneous (and likewise chimerical) forms of parenting, siblinghood, and kinship emerge where gender and sex are inflected by these arrangements.

My (Messy) Relationship to This Work

Doing interpretive qualitative research in this site of transgenic invention means juggling many dynamic and contingent elements of fieldwork. Investigating the topic and writing this book has forced me to keep many data points in mind simultaneously. I want to keep my head on straight, interpret the evidence through my positionality, thereby making it *my* project. I suppose I want to bring a certain order to a story that, like my life, isn't always orderly. For me, then, this book is and has been about living with the mess; I am gathering writing and still figuring things out about my past and present—I still live with and am now consolidating other writing and experiences about sperm donation, breastfeeding, leaking, mothering, intraspecies entanglements, transgender relationships. Plus, I'm now at menopause, which makes writing about an animal facing obsolescence pretty fucking personal.

I sorted out the story, recursively and continuously, and sometimes I got jumbled making sense of transgenic science, my confusion forcing me into a repetition of questions. The seemingly endless checking and rechecking of the correct order, quantities, techniques, and processes made me self-conscious with my human informants. I wanted to maintain my rapport (by not overburdening them), but I also wanted to appear smart, as though I could grasp this material as easily as they seemed to. And I suppose I also wanted them to like me. Furthermore, I now want my readers to like my informants (as I do).[6] Throughout my academic career and personal life, I've noticed that scientists can be easily demonized as collaborators in the medical- and military-industrial complexes. To do the science, the story goes, the scientists either deliberately, systemically, or automatically accept funding that corrupts their research. I myself have developed a type of interpersonal ethics in my methods so that even if scientists are wrapped up in some kind of capitalist military pharmaceutical drive, I do not know if they are any more wrapped up in it than I am. I want to commune with them the same way I want to commune with the goats or the spiders to enable another kind of companionate species knowledge to emerge.

Obviously, through my feminist science studies training, I am not seduced by a fantasy of objectivity. Yet my leaky self seems to get in the way, intent on being likable and truthful. I want an open conversation rather than the typical academic interchanges, which I find much more adversarial and critique-oriented than propositional and explorative. My work is sincerely driven by a ferocious curiosity and a desire for compassion. I reflect on my own shifting alliances from scientist to spider to goat to FDA regulator to arachnologist. I think about how research is conducted by us, fleshy beings connected or entangled in different ways. The connections can be warped by the politics of an always-gendered, and species-specific, knowledge production, and production in general, which includes stupid sociology, warring men, and human-centered domination of the rest of life, to say nothing of the faculty meetings.

Consider, for example, the herdsmen, these twenty-something college students. I wonder about their motivations beyond what they tell me. What can be gleaned by watching their social engagement with the goats? They are gentle with the goats but forceful too, securing the goats' heads in a metal device to milk them twice a day. I mimic their confidence with the does when I handle the goats, willing my hands to be as gentle as possible and trying to manage my own difficult emotions and sometimes equally difficult embodiment. As I walk the goats around the yard, they sense that I am a neophyte; they struggle and live up to their stubborn reputation. I don't think I am part of their herd or their tech. I imagine that they know something about me and are using it to their advantage—a bit more time on the walk, a sense of freedom, an act of resistance. And still perhaps my mom-ness can be an asset to knowing something about them. The goats reproduce and lactate, and I have done these things, so I recognize us as kindred. And in that witnessing, I come to know something of them. I must.

I try this angle out as I think and write. I ask myself, Is this reflexivity gone wild? Or perhaps it is reflexivity gone domestic (domesticated?). And what about pointing out that I am queer? In fact, am I using the term *queer* too much to describe myself? Especially because I don't really feel particularly queer, just anxious. I'm not writing a memoir, so what is the correct dose of inserting the self in the fieldwork, the writing, the thinking? Thinking back to graduate school in the early 1990s, I return to the previously mentioned cultural studies class with Jim Clifford. I recall him warning us not to insert too much of ourselves into our ethnographic work as it can appear lazy or gimmicky, thereby cheapening the scholarship. Identity characteristics don't explain anything or inoculate against bias without being carefully unpacked. In other words, just because I am queer doesn't mean that I am immune from some heteronormative bias. Nor am I released from the sustained and difficult work of unpacking heteropatriarchy. Certainly, my particular relationship with coordinate animal breeding and transgenic creation has been deeply imbricated by my distinctive positioning within histories of white supremacy, anti-Black racism, and settler colonialism. As a white highly educated woman growing up in the 1970s and 1980s in the United States, I've come to know and feel and emote in particularly raced, classed, and gendered ways. Furthermore, as discussed previously, in what ways does my white womanhood become affectively predictable, there for me to perform as I sort out how I feel about the goats or research and writing?

Is this too much? Working on this project has taught me different uses of language, often quite evocative language. *Freshening* and *drying off* describe stages of goat milk production; giving birth is *throwing kids*. As I have learned these words and processes and spent time with the spider goats, the humans, and the spiders, I have also come to see things differently, including seeing my own self differently. Just as the goat has become a system to express spider silk, I can now see my own body as a system. I've thrown three children through the use of donor insemination. To get pregnant, I've used two known donors and one identity-release sperm donor in an eleven-year span. Each time, I used a different man's semen, previously ejaculated into a collection jar and delivered to me on a couch or an exam table. Using a syringe and a speculum, another person (a partner, a doctor, or a husband, depending on the circumstances) injected semen on top of my cervix. And for nine months, I had dreams of meeting this new person.[7]

But now I also reimagine myself as a staging ground for an experiment of mixing my fluid with another fluid to produce a final result—it was a bit untamed science. I also reconsider my using a breast pump as akin to being mechanically milked. I think back to ways of being herded by various men in my life, directionally tapping me or steering me, their hand on my back. My connection to coordinated animal breeding and rearing of young has become more sturdy.

With some reluctance, I have realized that one thread of the spider goat story concerns a venture that has not yet yielded results. I hesitate and cringe a little as I write this statement, for fear that I am insulting my human informants. Clearly the Spider Silk Lab has produced volumes of scientific papers, presentations, laboratory protocols, and new types of animals; it has launched careers and cross-pollinated scientific studies in synthetic biology. These results are impressive and generative and not to be discounted. But still, these were not the results the investigators set out to obtain per se. There has yet to be a real tangible product constructed of spider silk protein currently manufactured and used. In some ways, this lack of practical utility makes the spider goats an extravagance. The spider goat is beautiful in its ingenuity, but increasingly she is deemed no longer useful, productive, affordable, or state-of-the-art. There has been fallout from this—the goats are made unnatural, and over time, I fear they will become the residual, the forgotten, something no one thinks about in the near future. I see this happening in my own life too: the same culture that would cruelly forget the spider goats can do the same to aging, no-longer-reproductive women. Hence the fantasy about being a desperado goat liberator, a bandana round my neck and SPF 50 to protect me from the sun.

Throughout this project, I've grappled with moving beyond a purely Marxist analysis, though it is very apt. It's fairly obvious that capitalist incentives have infiltrated and corrupted all knowledge pursuits in ways that prevent any quest for information outside a profit motive. People innovate these goats to make money. I wrote this book in part to maintain occupational prestige for continued success in my career. However, my book does expose the complex financial interrelationships and metapolitical economy that coproduce spider goats and their progeny. These hybrid goats are caught in a bureaucratic maze that connects two of our most powerful institutions—the military and medicine. The spider goats' silk protein is turned into safety, money, power, and salvation.

It has been suggested to me that this book might actually be my commentary on academia. In other words, I am using the transgenic goat to make a metacomment on academia and academic production. We take an object, a phenomenon, or a concept, then we modify it and milk the shit out of it and extract it and attempt to transmit it as a raw material for others to work with it. We do the best we can to make seemingly weird ideas profitable. And by profitable, I mean able to gain traction as a citable concept to increase the citation count, academic currency.

Just like the lab workers and the scientists, academics are also dedicated people working on projects that might not yield results. Spider silk research has provided livelihoods for decades, and even if we can't see the future for the herd of goats, the technology still has a life and created something that is not wholly nihilistic or wholly negative. We are complicit, all of us, and we can't escape the effects of chain reactions that capitalism fuels and that can cause harm at every corner of interaction. There is no way to avoid being tainted by it.

And still scholarly work marches on, attempting to be flexible in the face of the pitfalls of working with unstable systems. In the pursuit of spider silk protein at the Spider Silk Lab, a great deal of the synthetic biological innovation of all the products originated from the modification of goats. Justin summed up the different systems that have been modified to supply a steady and reliable form of spider silk protein:

> *E.coli* does have its shortcomings. But so do other hosts. Animals [goats] are expensive systems to work in and are rightfully subject to relatively intense oversight. Animal cells, grown in cell culture, are outlandishly

expensive [for commercializing] anything but a therapeutic form. Yeast are also relatively small and can have problems with getting rid of foreign DNA (spider silk, in this case). Plants make relatively large proteins and are very scalable, but purification of proteins from them is a challenge at scale due to chlorophyll and other contaminants. Silkworms are highly scalable, and an entire industry exists to grow the silkworms. The drawback, if it can be called that, is that the fiber is only about 35 percent spider silk protein. With that said, that leads to silkworm silk with spider silk mechanical properties, but it is not pure spider silk. However, of all these systems, silkworms are the only system that spins a fiber as part of the production process.

So, I would say there is no absolutely perfect system to produce these proteins. *E. coli* represents one of the best compromises for the small or midscale production of protein powder that can be used for a variety of different applications other than just fibers. Silkworms are pretty obviously the way to go if the desired output is a fiber.

Here I am again reminded, and perhaps a bit sadly so, that the goats themselves are really becoming something of the past. They are not seen as viable or scalable, and their gene line is petering out, the goats having served their purpose. While working on this book, I felt as if I was researching the coming transgenic future while also grappling with transgenic pasts. How can the spider goats themselves be critically endangered so shortly after they have been made?

My research on them has happened as the scientists themselves have been concluding that spider goats are not the future of spider silk production. I'm not sure if we can collectively mourn this type of animal extinction.[8] Is it legitimate to mourn a transgenic species as it probably goes extinct, the same ways we have been encouraged to mourn a natural species?[9] My boyfriend gets annoyed when I say I am sad to see the possible end of the spider goats. He wants to know why I am sad. I tell him it's because they are a form of life, and life can be grieved about.[10] I say that I have seen and met them and that they are unique. I'm attached to them in that they have hopefulness infused into their bodies despite their changing purpose. He is not convinced. "Many would argue that they never should have come into existence," he counters. I nod but I remain firm—they already exist, so they should be allowed to continue.

It gets a bit heated as he doesn't relent. He believes I am not sad for the goats but instead sad for the scientists for being "unsuccessful." He thinks my feelings are not for the goats at all but for the humans. Or if I am sad for the goats, he suggests, "Is it not because you have projected a human and attendant capitalist ideological wish onto them, to be productive, make something of themselves, be worthwhile, matter in terms of knowledge production, or product production? Why is it sad necessarily for a system or life to become something of the past?" And here is the crux of where we might differ. He seems to be saying that my attachment to the goats is inextricable from the circumstances in which they were created and that I take those circumstances as fact. And I do think that the spider goats themselves have value outside of what they do for us. I also wonder if this is an artifact of his being fifteen years younger than me; is my age an asset here?

But it is hard for me to articulate why—other than to say I just feel it. Questions of attachment have circulated in this book—attachment to heteronormativity, to progeny, to being pure, to being contaminated, to conservation (saving the goat), and, by extension, to being alive. In the broader frame of geological time, an attachment to a particular species (or subspecies) is a bit trivial. Other events consume that small event. Yes, the Anthropocene means that many things are happening because of us, but thinking of turning anthropogenic change around is just as arrogant.

Obviously the spider goats were created by and for us. But their life must have other worth. They were treated as a system, a technology, and since technology is constantly changing and improving so quickly, the writing was on the wall for these goats at their conception. It was foretold that they would be surpassed. Although they were the vehicle by which we can now do other things in biomaterials using spider silk, it feels somehow wrong to declare, "You did your job, and now we don't need you anymore." Humans can generate a unique, transgenic species and then extinguish it at will. For some humans, the spider goats have lived out their purpose. We have learned everything we can from this animal, and now it's time to put them away—by euthanizing them or letting them die out. They may soon be an extinct technologically constructed species that came and went in my lifetime, but they are not a natural species. So perhaps that means their extinction wouldn't be as tragic.

But it's not as if I have a well-thought-out plan of what to do with them. The whole time I have been working on this book—from the minute I started mentioning spider goats until this very second—everyone I have spoken with eventually says, "Well, what should we do about them?" Release them to an unpopulated island somewhere and ruin that ecology? Sterilize them and watch them age, but keep some frozen embryos for potential regeneration at some point? Create a sanctuary on some rich, goat-loving human's sprawling ranch for their social media brand? There is a steady pressure to be certain and firm—to make assertions such as the goats are being exploited or the science is good or bad—but I am resistant to doing this type of work for the reader.

Science fiction can offer some options. Margaret Atwood's *MaddAddam* trilogy crafts a world of biologically engineered genetic hybrids within the backdrop of corporate domination over everyday life amid ecological degradation.[11] The characters facing human extinction must also address the introduction of a new race of beings, Crakers, and innovated transspecies, which include "pigoons," "wolvogs," and "liobams." In the first book, Atwood introduces spider goats, which she calls "spoat/gider." These creatures roam and challenge other species, including humans. A new ecosystem evolves, and it includes all kinds of humans.

Arguing with people, including imaginary readers, throughout this project, I see how I am caught up in a sweep of emotions, entangled in a situation. The lurking question of my relationship with my biotech children remains. What standards could or should be applied to our relationship as humans with our living commercial biotech (my children, the goats)? Maybe I was asking these types of questions because I am a woman (who likes fluffy goats), but it's also because I am a mother, likewise a material-semiotic assemblage shaped by technology, capitalism, and planned obsolescence.

Humans, Goats, and Spiders in Contact

My exploration of the goats and spiders has led to different locales and situations, including laboratories and farms in Logan, Utah; roadside forest brush in Gainesville, Florida; hallowed museums in New York City; and interviews with a Slovenian arachnologist, a California textile company founder, and a North Carolinian goat breed authenticator, among others. This geographical spread and motley collection of places, people, and things has compelled me to think about myself and my own family's motley makeup. The production of my children is an assemblage of beings, materials, places, technologies, affects. I yearn for a wholesomeness that, paradoxically, also repulses me, in part because it has shunned, and continues to shun, me. I face the world with an anxious, queer suspicion that fills me with pessimism every single day. I have experienced an internal, irrepressible electricity when jumping over fallen trees and spotting a spider in the center of her web backlit with sunlight. My babies, my own children, are approaching expertise in the field of biogenetics and, in a role reversal I thought (wished) I was too young to experience, are teaching me.

Goats have been co-constituted with humans as domesticated species for centuries. Humans also use goats as symbols in mythology, as agricultural tools, as systems for medical treatments, as food, and as clothing material. We have raised each other up—we breed them in ways that work for us, and they enable us to wander and roam. Spiders precede humans on the planet. Like many other invertebrates, spiders are considered weird, alien, and creepy. Some psychoanalysts see spider phobias as related to the mother—at once nurturing and devouring. Though goats are more mammalian, domesticated, and social animals that live in communities with humans, we have deep connections to spiders as well. Humans have put these two animals together—mixing the vertebrate with the invertebrate, the mammal with the arachnid, the domestic with the wild, the herbivore with the carnivore. We now have multiple generations of self-producing spider goats. How do humans reckon with creating a wholly dependent species that must be managed and maintained?

Breeding Designer Animals

In conducting my fieldwork, I've definitely had the experience of enjoying things that I am supposed to deconstruct as truly exploitative. Unparalleled pleasure flowed in capturing those spiders in Florida—the sense of danger, accomplishment, beauty, and contact. And yet, I was participating in a practice of ripping a species from its habitat to ship in a plastic container to a new location for the entertainment of privileged American children. I see myself as performatively stifling my own delight to get down to the business of critical analysis. Ethnographers who study Walmart workers aren't supposed to proclaim, "Gosh! That place is so spacious, with great lighting and *anything you could ever want or need*! It's astonishingly beautiful in its abundance."

This paradox is akin to how poverty appropriation can be kind of a rite of passage in sociological circles—from observations of undergraduates' interactions to my own collegial relationships. It's the sense that "enlightened" people think they aren't supposed to like bourgeois things or have such desires, as if what you want or don't want or how you feel is just a pure reflection of your class location and reveals your false consciousness.

In the end, what I know is this: I really liked being with spider goats and enjoyed meeting them. I don't think it is all bad that they were made. As I mentioned in the preface, I believe that some of the social worlds in which I circulate—sociologists, moms, lefties, progressives—expect me to be deeply troubled about the making of the goats. When I talk about the project at dinner parties, the dog park, the yoga studio, or family gatherings, people are usually both fascinated and disgusted. They ask me, "How could they do that to the goat?" and "Isn't that illegal?" They tell me, "That goes against nature."

It is almost comical to hear these remarks at the dog park, from someone throwing a ball to their purebred labradoodle or the owner of a large English bulldog snuffling over a stick. They see little connection between the making of their own companion animals and the creation of the spider goats. But quite ironically during an interview, Randy practically anticipated this very experience, saying, "Take a look at dogs right now. What have people done with dogs? Everything, absolutely everything. Has anyone said a word? Not one word from anyone. So, we have been manipulating for several thousand years what we want in dogs, and we end up with arguably dogs that are completely and totally nonfunctional without a human being to care for them."

To date, there are 195 breeds of dogs recognized by the American Kennel Club.[12] Contemporary debates, as featured in the news media such as the *New York Times*, cite the high frequency of inherited health problems, including eye ulcers, breathing problems, and hip dysplasia, in purebred animals with limited genetic diversity.[13]

Danielle Gasinby, the graduate student from Randy's lab, added her own take on some people's criticism of genetically modified animals:

> Goats . . . we have had problems with PETA [People for the Ethical Treatment of Animals] and other groups. Our goats—there is nothing different about them; they make one more protein extra in their milk. We jump through all these hoops for the goats. No one cares about the spiders, and no one cares about the bacteria. All of our drugs come from genetically modified bacteria, and no one cares about that. I think if people really understood the process, maybe they could be more informed. But they think the treatment of the goats is cruel. Their manipulation happens as an embryo. We aren't torturing goats. No one cares if you dissect a spider; we don't like spiders. We like goats, though.

Danielle is articulating something that I have also experienced repeatedly during my work. I have conducted extensive research with, and written books about, humans, bees, horseshoe crabs, goats, and spiders. Notably, it's fairly predictable how people rank the hierarchical value structure for animals when I share my scholarly trajectory.[14] Clearly, we are on top, followed by goats, then bees, then crabs, then spiders. The level of concern over human intervention with each animal decreases as we move further away from our own taxonomic classification. Humans don't love all animals equally: some animals' circumstances become "deserving" of some compassion or inquiry, while others' exploitation doesn't even register in human thoughts.

My lived experiences inform my increasingly fragile attachment to naturalness, exposing cracks in the facade of my claim (and others'?) to biological integrity. I have no personal experience with the natural way of reproduction. Although I have given birth to three children, I have exclusively reproduced through donor insemination. The so-called natural way of having children is a phantom. I am haunted by the natural in my self-concept (Am I a real mom?), my ways of explaining my family to others (How are we related to these children, and how are they related to one another?), and the origin stories I tell my children (Where did they come from?). In turn, the way my children understand themselves invokes their phantom "real" fathers. As I attempt to be a sociologist, I reflect on my own experience and my daughters' experience of being haunted by naturalness, and I become more sympathetic to scientific mediations of biological reproduction.

As I was putting the finishing touches on this book, all of us humans were facing a great many contradictory, intense, multifaceted, and changing reactions toward the natural world brought to the surface by a raging and evolving pandemic. The mRNA vaccines come to mind as they represent a new technology in the production of vaccines. Vaccines are injected into the human body as a way to activate the immune system and generate antibodies.

An mRNA vaccine uses, as an antigen or a foreign body, mRNA (messenger RNA) instead of the traditional dead or live-attenuated virus (a virus whose pathogenic properties have been drastically diminished) or even subunits of a virus such as viral proteins. For the COVID-19 vaccine, the injected mRNA carries the instructions on how to build a spike protein that mimics the spike protein of the coronavirus. The body, reacting to this protein as if it were the actual coronavirus, makes antibodies to fight this "invader." Although the mRNA quickly breaks down in the human cell, the cell's ability to marshal its ability to create antibodies remains. If a vaccinated person is later exposed to the real coronavirus, the body now "knows" how to produce the antibodies to deactivate the virus. For better immunity against COVID-19, we need to have this mRNA sequence injected into our bodies, because humans don't naturally have the instructions to code for the spike protein. In the same way that the goats don't naturally produce spider silk proteins, we lack the mRNA that codes for this spike protein.

The pharmaceutical industry's production of these mRNA vaccines is called *synthetic* because the viral mRNA production methods are not natural. Large quantities of the viral mRNA can be produced chemically, rather than through biological synthesis, in an in vitro synthetic system. Through *in vitro synthesis*, a product is made outside a living organism. The spider goats are an *in vivo synthetic system*; the spider silk protein is produced inside their living cells and bodies.

Initially, the mRNA vaccines were treated with more caution by institutional actors and investors, including big institutional buyers like the European Union (with far more intense hostility expounded by conspiracy theorists and religious fanatics). mRNA vaccines in humans were just theoretical until this COVID-19 vaccine. This vaccine is the first mRNA to be approved by the FDA for widespread use. The controversy exists partly because mRNA vaccines had not been thoroughly tested in humans before this vaccine rollout (a US effort called Operation Warp Speed) so there was little known about its long-term effects. Even the pope felt compelled to make a statement of approval of these particular vaccines because although they have some connection to stem cell research, their benefits outweigh their controversial beginnings. And yet, in just a few months, and as the superiority of this technology in terms of safety, effectiveness, and adaptability to mutations has been shown, much of that cultural resistance has subsided. It will be interesting to watch how we will react to leaps in treatments and vaccinations that will certainly emerge after this crisis is over because of this technology. mRNA vaccines are easy and fast to produce, so they could be used in future outbreaks of viral infections. Scientists have to manage these innovations in the face of public anxiety about the natural world and shifting risk-assessment strategies.

Scientists, popular opinion sometimes goes, use unnatural ways of intervening in "natural" processes and create monstrous freaks of nature—spider goats, for example. But I find myself shifting my attachments (and allegiances) away from the natural to the freaks, the

Figure c.1. Walking a transgenic goat back to her pen from the milking barn. Photo by C. Ray Borck.

transgenes, the mediated deliberate blends, even perhaps seeing myself, in retrospect, as creating them.

We humans are continually reckoning with our relationship with nature. On the one hand, we situate ourselves outside and above the natural world to justify (or at least feel OK about) the exploitation of nonhuman animals and natural resources for our own needs and wants. On the other hand, we fall back on the myth of the natural when it serves other purposes—such as explaining (probably socially constructed) behaviors and ways of being. As we swivel between these and other understandings of ourselves, the blurry spaces in between are left mostly unexamined. And our (human?) desire for easy categorization, explanation, and differentiation can't be satisfied when it comes to scientific research, transgenic goats, human insemination, mechanized milking, and purebred dogs. Nor can it probably be satisfied in any exploration of the natural world, much as we wish it could be. Like the goats, these ideas are hard to corral. They weave and reweave in motion.

ACKNOWLEDGMENTS

These four years of studying and writing about spider silk, spiders, goats, and scientists have been made possible through kindness, patience, and collaboration. I have been extraordinarily lucky to be welcomed into the Spider Silk Lab at Utah State University. This research would not have been possible without Randy Lewis and Justin Jones, two individuals who have been unfailingly honest, accessible, and open about the work they do and the ethic that guides their devotion to science. The members of the USU team, including Amber Thornton, Brianne Bell, Brittany Grob, Andrew Jones, Thomas Harris, Xaoli Zhang, and Christian Iverson, have each been generous, kind, and available to me throughout this project. My good fortune at meeting these individuals has changed my life.

The scientists Cheryl Hayashi, David Breslauer, Matjaz Kuntner, Lawrence (Lary) E. Reeves, G. B. Edwards, Alison Van Eenennaam, and Larisa Rudenko each taught me about scientific processes, animals, synthetic biology, federal regulations and guidelines, and transgenics. My research is indebted to them.

My colleagues at Purchase College, the anthropologists Jason Pine and Shaka McGlotten, have read this manuscript and offered keen insight and loving support. The previous chair of my division, Linda Bastone, enhances my life at Purchase and always has the best candy. The Sociology Department, including Chrys Ingraham, Kristen Karlberg, Matthew Immergut, Mary Kosut, Alexis Silver, and Toivo Asheeke, continue to make Purchase a remarkable place. In particular, my friends Matthew Immergut and most especially Mary Kosut have provided me with their smart reading and generous support on this project and throughout my career at Purchase. Special thanks to additional colleagues who helped to clarify particular points: Steve Flusberg, Stephen Cooke, and Paul Siegel. Special thanks also to Troy Vettese for early chapter reading.

Darcy Gervasio continues to be so ingenious in her research skills and literary insights that I am always grateful and awed. My current and former students breathe life into my scholarship. In particular, Anaïs Baptiste, Brendan Regan, Anna Krol, Andrea Farley Shimota, and Heidi Durkin carefully read and commented on various stages of this manuscript. I thank them for their time and dedication. Gratitude to Grace Moore's organizational skills and patience in collating the images. Special thanks to Nina Lorcini of Dúagwyn Farm and her goat Saxon for the author photo.

The Bedford Hills College Program students enrolled in Critical Animal Studies during spring 2020 provided smart and relevant feedback on earlier stages of this manuscript.

I don't know how I got to have so many smart and vibrant friends, but I am so deeply inspired by them and their willingness to read different drafts of this book, especially Monica Casper, Tine Pahl, Maria Sereti, Karin Schott, Matthew Schmidt, Veronica Kaleta, Marni Corbett, Patti Curtis, Patty Howells, and Shari Colburn. Grateful to Matt especially for being so genius.

I have benefited from the financial support of a grant from my union, United University Professionals through the Joint Labor-Management Committee and the Provost's Office at Purchase College, State University of New York.

Several years ago, my friend Megan Davidson and I started a writing group; our friendship has flourished, and our creative work is juicier. This book would not have happened if it weren't for Megan suggesting the topic; I am forever grateful to her, C. Ray Borck, Jenni Quilter, and Stephanie Schiavenato for how they push me at the precise moments, soothe me when I'm spiraling, and indulge me in sometimes-tedious conversations to gain clarity. I am humbled by their intelligence and grateful for their persistence.

For earlier commentary on this book I thank Larin McLaughlin, Banu Subramaniam, and Rebecca Herzig. I am also grateful to the members of the workshop Animals on the Left, a group organized by Troy Vettese, in particular Luisa Reis Castro and Aaron Van Neste.

For more than seventeen years, Ilene Kalish has been honest, trustworthy, smart, loyal, and reliable. I am forever indebted to her for all the ways she has contributed to my career and my life. I am also thankful for Sonia Tsuruoka, Martin Coleman, Patricia Boyd, and New York University Press.

Finally, my parents, Linda and Richard Moore, have fostered in me a deep curiosity about the world and for more than fifty years created plenty of adventures for subsequent and ongoing investigations. My aunt and godmother, Joan Pendergast, constantly buoys me with reminders that I am her favorite niece. Thank you to my extended family members Paisley Currah and Robyn Mierzwa.

My remarkable children, Grace, Georgia, and Greta, are infused in everything I do, everything I think, and all that I write. I am astonished at who they are and who they become each day. Thank you for being my kids and letting me be your mom. And my partner, C. Ray Borck, makes everything worth it by creating joy in the most unexpected moments and infecting everything with divine perversion. This book is dedicated to him.

NOTES

PREFACE

1. Hekmat and Dawson, "Students' Knowledge and Attitudes."
2. See my previous collaborative work, Moore and Schmidt, "On the Construction of Male Differences."
3. Moore, "Incongruent Bodies."
4. Paisley Currah, telephone conversation with author, October 10, 2020.
5. Haraway and Goodeve, *How Like a Leaf*.
6. This brings to mind the anthropologist Claude Lévi-Straus, *The Savage Mind*, who proposes a theory of binary opposites to explain mythology, signs, and symbols, and he strikes upon the particular comparison of the bricoleur versus engineer. A DIY tinkerer or a bricoleur makes do with whatever is readily available to handcraft something new, something that would not be expected of the raw materials. There is a playfulness, naughtiness and originality in the bricoleur's work. The bricoleur is the maker of myths. In contrast, the engineer, who is more of a skilled professional and works to create a stable system. There is sturdiness, stability and rigidity to the work of the engineer. This engineer is the maker of Western scientific discourse. Like the bricoleur I have made these unnatural children but the engineers, the scientists, are in the labs making more fantastic creatures. Gifts from science.
7. Whatmore, "Materialist Returns."
8. Gruen and Weil, "Animal Others."
9. Gruen, *Entangled Empathy*.
10. Bloch, "Community Embedded in Mobility."
11. Joshi, "Indian Domestic Goats."
12. Singh, "Migratory Sheep and Goat Production System."
13. Choudhary, "Pastoralists of Himachal Pradesh."
14. Haraway, *Modest_Witness@Second_Millennium.FemaleMan©_Meets_OncoMouse™*.

INTRODUCTION

1. Franklin, *Dolly Mixtures*.
2. Blanchette, *Porkopolis*.
3. Sunder Rajan, *Lively Capital*.
4. For the documentary about the Anthropocene as it affects species living on the shoreline, see Sean Hanley's website at www.sean-hanley.com. See also Hanley, "The Whelming Sea."

5   Hartigan, *Shaving the Beasts*, 18.
6   High, "Playing with Rats."
7   Taylor, *Beasts of Burden*.
8   Taylor, *Beasts of Burden*, 8.
9   Nelson, *Social Life of DNA*.
10 Benjamin, *People's Science*.
11 Pollock, *Synthesizing Hope*.
12 Roy, "Feminist Practices for the Natural Sciences."
13 Barad, *Meeting the Universe Halfway*.
14 Yanagisako et al. *Essays in Feminist Cultural Analysis*.
15 Adams, *Sexual Politics of Meat*.
16 Weinbaum, *Afterlife of Reproductive Slavery*.
17 Hill Collins, "Sociological Significance of Black Feminist Thought."
18 Jackson, *Becoming Human*.
19 In a powerful example of how so-called human prerogatives made accessible to Black women do not result in greater equity, Jackson explains that when the black female body gains access to forms of class ascendancy or education, black reproductive health outcomes actually worsen. She goes on to show how Black people who enter white-majority locations and are immersed in whiteness can produce a type of debility.
20 Interview from Crandall, "Zakiyyah Iman Jackson on Becoming Human."
21 Wang et al., "Recombinant Human Lactoferrin"; Taussig, "Bovine Abominations"; Lewis, "New Kind of Fish Story."
22 Clark, "Killing the Enviropigs."
23 Scott, Inbar, and Rozin, "Moral Opposition to Genetically Modified Food."
24 Douglas, *Purity and Danger*.
25 Scott et al., "Attitudes toward Genetically Engineered Food."
26 Lusk et al., "Genetically Modified Food Valuation Studies."
27 Scott et al., "Attitudes toward Genetically Engineered Food."
28 Öz, Unsal, and Movassaghi, "Consumer Attitudes toward Genetically Modified Food."
29 Scott, Inbar, and Rozin, "Moral Opposition to Genetically Modified Food."
30 Runge et al., "Attitudes about Food."
31 George Freeman, quoted in Hennessey, "GMO Food Debate in the National Spotlight."
32 Beck, *Risk Society*; Giddens, *Modernity and Self-Identity*.
33 Giddens and Pierson, *Conversations with Anthony Giddens*, 27.
34 Giddens, *Renewal of Social Democracy*.
35 Ekberg, "Parameters of the Risk Society," 346.
36 Laurent-Simpson and Lo, "Risk Society Online.
37 Stengers, *Cosmopolitics*, 9–10.
38 Grounded theory is a deductive process whereby analysts incorporate as much data as possible to use the formative theories as deductive tools. Through the

writing and rewriting of analytic memos, this tool, the grounded theory, ultimately aims to incorporate the range of human experiences in its articulation and execution. According to Anselm Strauss, a key developer of grounded theory, it is through one's immersion in the data that these comparisons become the stepping-stones for formal theories of patterns of action and interaction between and among various types of data. By triangulating data sources about spider silk and goat representations over the past three years (scientific texts, field notes, interview transcriptions), I established various points of comparison to explore multiple concepts about silk, goats, spiders, transgenics, and so on. Working with my analytic memos written about these concepts, I established interrelationships between them. "Theory evolves during actual research, and it does this through continuous interplay between analysis and data collection" (Strauss and Corbin, *Basics of Qualitative Research*, 273). Similar to other qualitative research, content analysis can be *exploratory* and *descriptive*, enabling limited insight into why significant relationships or trends occur. The aim is not toward standardization of facts into scientific units but rather an appreciation and exploration of the range of variation in a particular phenomenon. Outliers, representations that do not fall neatly into the collection of the most common themes and concepts, are useful because they enable analysts to capture this range of variation and dimensions of the concepts.

39 See for example, Billo and Hiemstra, "Mediating Messiness."
40 See, for example, Joseph, "Relationality and Ethnographic Subjectivity"; Banister, "Women's Midlife Experience"; Davies, *Synthetic Biology*; and the special issue devoted to the topic, Davids and Willemse, "Embodied Engagements."
41 Considering the classic sociological text Riesman, Glazer, and Denney, *Lonely Crowd*, and the preoccupation with understanding the nature of conformity, I can see how analysts can elide the nuances of resistances and rebellion.
42 Justin Jones, one of the principal investigators of the lab, said, "At least yet with the current state of technology, we can't reproduce what a spider does, except for using another biological system. So we're going to make spider silk from goats, we're going to make it from bacteria, from alfalfa, from silkworms, and we are going to continue to modify these systems to make the protein and create applications for the protein."
43 A redux, where biological engineers appropriate and repurpose tools to solve mechanical problems, comes to mind. See Clarke and, *Right Tools for the Job*.
44 See Animal Planet, "Spiders," season 1, episode 1, 2021, of *Best Kept Secrets*, www.animalplanet.com.
45 Indeed, *herdsman* is a gendered term, but it is the term used routinely in the lab. It means the owner or the keeper of a herd of domesticated animals, and although it is gendered male, it is used as the universal term to refer to any human who take care of the goats.
46 I was in Logan in August 2017 and July 2019.
47 Deleuze and Guattari, *A Thousand Plateaus*.

48 Moore and Kosut, *Buzz: Urban*; Moore, *Catch and Release*.
49 Through Hayashi's connections, I also spoke with David Breslauer, chief scientific officer and cofounder of Bolt Threads, a materials innovation company that uses the principles of *biomimetics*, the human-made processes, substances, devices, or systems that imitate nature to develop and create new textiles and other materials that raise the bar for sustainability. The tone of the interview was very friendly with a degree of polish that indicated David was adept at speaking with the press. I also interviewed Matjaz Kuntner, an arachnologist from Slovenia and a leading expert on golden orb weaver spiders. We spoke over Skype just as he was preparing to go on a collection trip in Asia. Chapter 1 describes my fieldwork conducted in Gainesville, Florida, during an expedition to collect golden orb weavers. Having spent time in wooded roadside areas with spiders and having interviewed two thoughtful entomologists, I describe these experiences in detail as a way to examine spiders' silent participation in spider silk science. My style of interviewing is deeply interactive. I record all interviews, transcribe them, and code them for themes and patterns. I also video record many lab and fieldwork procedures and watch and rewatch them multiple times to reflect on scientific practice and animal husbandry. My qualitative method also relies on respondent validation as a means of verifying my observations and explanations of scientific processes. Whenever possible, I check my written interpretations with my informants through emails and follow-up phone calls.
50 Moore et al., "Paleo-metagenomics of North American Fossil Packrat Middens."
51 Goldstein and Johnson, "Biomimicry."

1. SPIDER ENCOUNTERS

1 White and Williams, *Charlotte's Web*.
2 Takamoto and Nichols, *Charlotte's Web*.
3 Spiders appear as fictionalized characters in many children's and young adult literature beyond "The Itsy Bitsy Spider" and "Little Miss Muffet." Heads up to my librarian Darcy Gervasio for pointing out the spider from *James and the Giant Peach* and Shelob the giant spider from *The Lord of the Rings*. Myths animate spiders cross-culturally from the Ancient Greek Arachne, the maker of tapestries, to the West African folktale of the trickster Anansi. See Punter, "Arachnographia: On Spidery Writing."
4 While Charlotte is not a golden orb weaver (*Trichonephila clavipes*), as a barn spider, she is a member of the orb weavers (*Araneus cavaticus*).
5 Wilkie, Moore, and Molloy, "Anthrozoos and Society & Animals."
6 Both horseshoe crabs and spiders are in the phylum Arthropoda. As arthropods, the two types of invertebrates evolved from marine segmented worms, and both have exoskeletons, segmented bodies, and jointed appendages. Even more specifically, horseshoe crabs are more related to spiders than other crabs are, because spiders and horseshoe crabs belong to the same subphylum, Chelicerata, meaning they have chelicerae, or jawlike pinchers, in front of their mouth parts. Another

similarity is that horseshoe crabs and spiders also molt to grow. The two animals' anatomies are similar. They have book gills (crabs) or book lungs (spiders). Male spiders have a pair of appendages (pedipalps, or "boxing gloves") that are useful in reproduction, and male horseshoe crabs' first set of legs are also called boxing gloves and are used to attach to the female at reproduction.
7 Moore and Wilkie, "Introduction to *The Silent Majority*."
8 Atkinson, *Tattooed*.
9 This statistic comes from an oft-cited German study of arachnophobia: Schmitt and Muri, "Neurobiologie Der Spinnenphobie. Schweizer."
10 Hoffman et al., "'Spidey Can.'"
11 Dalton, *Spiders: The Ultimate Predators*.
12 In email exchanges with me, entomologist Lary Reeves shared several observations about spiders: "Bolas spiders produce fake moth pheromones. They mimic the smell of female moths, luring in males. When the males fly in, they swing that sticky glob of silk at the moth, essentially lassoing it to pull it in. Net-casting spiders build a silk net, and hang around above a spot where they expect an insect to walk, then slam it down onto the unsuspecting insect. Spiders in the genus Naatlo, and a few others, build kind of a standard orb web, but sitting in the center, they attach an anchor line and stretch the web backward along the line creating tension. When an insect flies in front of the web, the spider releases the tension and the web flies forward. Some species of Cyclosa build structures that resemble larger spiders at the center of their orb webs. No one knows the function of these, but one idea we had when working on these in the Amazon is that they help protect the spiders from spider-eating helicopter damselflies that specialize on Cyclosa-size spiders, but avoid larger spiders. No one really knows anything about silk henge [a minuscule Stonehenge-like structure containing spider eggs in the center and silk "stones" or fencelike shapes surrounding the eggs] other than it's created by a spider to protect an egg. But, it certainly is an interesting example of how spiders use silk."
13 Cheryl Hayashi told me, "If you count the glue, which I do, then *Nephila clavipes* have seven [types of silk], which are major ampullate, minor ampullate, flagelliform, cylindrical, aciniform, piriform, and aggregate."
14 The spinneret, an organ typically located on a spider's abdomen, winds fibroin protein molecule threads into a solid but sticky gossamer.
15 Saravanan, "Spider Silk."
16 White and Williams, *Charlotte's Web*, 179.
17 Roos, *Correspondence of Dr. Martin Lister*, 48: 202.
18 Roos, *Martin Lister and His Remarkable Daughters*.
19 Levi, *Spiders and Their Kin*.
20 Ouellette, "Wonders of Spider Silk."
21 Soth, "Tangled History."
22 Ingold's essay challenges actor network theory through a story of an ant and a spider in a discussion about networks and agency. Ingold's SPIDER argues that agency is not equally distributed between all organisms and their meshes (the

material conditions in which they live) and comes from growing and developing a skill. Ingold, "When ANT Meets SPIDER."

23 Weber, *Web Watching*.
24 M. Kuntner, E. Rudolf, and P. Cardoso, "*Nephila clavipes*," *IUCN Red List of Threatened Species* (Cambridge, UK: International Union for Conservation of Nature, 2017), www.iucnredlist.org.
25 C. Linnaeus, "Araneae," in *Systema Naturae* (Salvius: Holmiae, 1767), 1030–1037.
26 Joanne Randolph, *Orb-Weaver Spiders*, Nightmare Creatures: Spiders! (New York: PowerKids Press, 2014).
27 Kuntner et al., "Golden Orbweavers Ignore Biological Rules."
28 Bordo, *Unbearable Weight*.
29 Nowak, "11 Monogamous Animals."
30 Chris, *Watching Wildlife*.
31 The laboratory that is part of the University of Florida is set up to study biting insects and vector-borne diseases and pathogens in Florida, with applications for global public health.
32 Exploring the concept of interstices in suburban areas is attributed to my conversations with Stéphane Tonnelat. For an example of his work, see Tonnelat, "'Out of Frame.'"
33 See, for instance, Grazian, *American Zoo*.
34 Kalof et al., "Fostering Kinship with Animals."
35 Whitley, Kalof, and Flach, "Animal Portraiture."
36 Herek and Glunt, "Heterosexuals' Attitudes toward Gay Men"; Burke et al., "Do Contact and Empathy Mitigate Bias?; DellaPosta, "Gay Acquaintanceship and Attitudes."
37 Allport, *Nature of Prejudice*.
38 G. B. Edwards was a wealth of information. At the end of the interview, I asked if there was anything else I should know about golden orb weavers. He replied, "Well, there's one off-the-wall fact that you might be interested in. If you ever get lost in the woods and there is nothing to eat, they're edible. I mean, the whole . . . most of the abdomen is just ovaries, it's all just egg. So you could just grab one out of the web and bite the abdomen and eat a few. There was a guy in England, William Burstow. He used to eat spiders to see what they tasted like. And I believe he said they tasted like peanuts or something."
39 Hochschild and Machung, *Second Shift*.
40 Zemlin, "Mechanical Behavior of Spider Silks." Thanks to Darcy Gervasio.
41 Zemlin, "Mechanical Behavior of Spider Silks," ii.
42 Zemlin, "Mechanical Behavior of Spider Silks," 14.
43 Carbone, "Pain in Laboratory Animals."
44 Phillips, "Savages, Drunks, and Lab Animals," 77.
45 As anthropologist Lesley Sharp's extensive work with humans working in animal labs attests, "lab personnel employ an array of strategies to foreground the moral complexities of their work and lives." Sharp, "Animal Research Unbound," 76.

46 Bishop, "Army Genetic Scientist."
47 Leary, "Science Takes a Lesson from Nature."
48 When Justin Jones was reviewing this information with me, he added, "This is generally true, that *E. coli* can only make smaller versions of the silk. There is one paper out there (maybe two from the same group) that claims to have made *E. coli* produce native-sized-ish proteins. No one has ever been able to duplicate the work, and the authors have not made their material (vectors, etc.) available to other researchers, to my knowledge." After successfully modifying the goats, getting *E. coli* to work making spider silk protein has become perhaps the biggest quest in the Spider Silk Lab. Justin said, "However, this is the point of genetic/biological engineering. How do we get them to produce larger proteins than what they have evolved to produce? Our group, as well as others, have been successful at this to some degree. We have produced proteins in *E. coli* in the range half the size of native proteins. The problem is that expression levels of the proteins fall off markedly the larger the proteins get. That becomes *the* problem with commercializing from *E. coli*." In other words, *E. coli* can produce the spider silk protein, but the amount of protein made is considerably less when the protein is large. This is not a problem with the goats.

2. THE GIFTS OF GOATS

1 My midwives said that the microbes in the guts of babies fed goat milk instead of cow milk more resembled the microbes in babies fed human breast milk. They also said that goat milk is easier for babies to digest. I am not endorsing my midwives' claims, however, as I have now learned that goat milk is not recommended for babies younger than one year.
2 Anne Mendelson, a food writer, points out that milk is mammals' first food. Mendelson, *Surprising Story of Milk*.
3 See, for example, Dowden, *Zeus*. Also, my children have been fascinated by the classic book d'Aulaire and d'Aulaire, *Book of Greek Myths*, which recounts the story of the goat fairy.
4 Doula Megan Davidson, interview with author, September 15, 2018, said that the well-known midwife Ina May is fond of calling forth instances of humans feeding nonhuman animals with their milk. Megan's own sister-in-law fed her own breast milk to a baby goat on her farm.
5 Radbill, "Animals in Infant Feeding"; Cohen, "Interspecies Right to Breastfeed."
6 Discussing ancient civilizations, Radbill, "Animals in Infant Feeding," 26, writes, "Not only were children nursed by animals, but animals were nursed by women. This was done for many reasons: to feed young animals; to relieve the women's engorged breasts; to prevent conception; to promote lactation; to develop good nipples, and for other health reasons. The custom occurred among the ancient Romans and Persians."
7 Simoons and Baldwin, "Breast-Feeding of Animals by Women."
8 For companion animals, see Haraway, *Companion Species Manifesto*. Regarding despised animals, Nagy and Johnson, *Trash Animals*, claim that trash animals

disrupt human enterprise and have little or no perceivable and immediate economic value to humans. These nonhuman animals are often cast as draining economic resources through their alleged destructive behaviors. Trash animals also enable humans to occupy a self-appointed role of ecological stewards without the self-reflection of our own behaviors. In assigning the label *trash* to an animal, humans assign negative value, justify violence often through extermination, engage in eugenic practices of forced sterilization or management of female bodies, and elevate their own human status as preeminent.

9 See the news story Putzier, "Landscaping Goats."
10 Like spiders, goats also appear in children's literature, for example, the lead characters in "Three Billy Goats Gruff" and Mr. Tumnus in *The Chronicles of Narnia*. Librarian Darcy Gervasio conducted a search on WorldCat (www.worldcat.org) and found approximately 393 titles of children's books (juvenile fiction) featuring goats and held at libraries around the world. She suggested taking these results with a grain of salt—some are picture books featuring goats as one among many farm animals. That said, goats do seem to feature heavily in English-language children's books and in folk tales, and there seem to be goat stories from many countries around the world, including Romania, Sweden, Italy, Ethiopia, Zimbabwe, New Zealand, and Japan, and goats appear in Puerto Rican, African, and Jewish stories.
11 Starbard, *Dairy Goat Handbook*.
12 US Department of Agriculture "Sheep and Goats."
13 Amundson, *How to Raise Goats*.
14 Rosenberg, "Charity Heifer International Does More Harm than Good."
15 Rosenberg, "Live Animal Charities Are Ill Conceived."
16 There are alternative renderings such as Sigil of Baphomet, the goat pentagram symbol as the official insignia of the Church of Satan.
17 Arluke and Sanders, *Regarding Animals*.
18 Amundson, *How to Raise Goats*.
19 For a peek into the goat dairy's operation, see Meyenberg, "About Our Goats," accessed September 15, 2021, https://meyenberg.com.
20 The terms *sire*, for the male parent, and *dam*, for the female parent, are used in breeding contexts with goats and other animals, such as horses and dogs.
21 *Freshening* describes the period when an animal begins to produce milk. After a doe kids (gives birth), she freshens; her milk comes in from giving birth.
22 On the does, the ear tags indicate the generation that they were born and the birth order in that generation. The radio-frequency identification (RFID) implants inserted behind the ears provide permanent ID for the goats.
23 Thanks to Lindsay Parme for this suggestion.
24 Schuller, *Biopolitics of Feeling*.
25 Schuller, *Biopolitics of Feeling*, 7.
26 For an example of this attachment versus detachment and engagement versus distance with animals in the field, see Candea, "'Carlos the Meerkat.'"

27 King, *How Animals Grieve*.
28 Anthes, *Frankenstein's Cat*.
29 Clarke and Fujimura, *Right Tools for the Job*.
30 Fish, *Living Factories*.
31 There is a history of using goats in military research. According to Linzey and Tutu, *Global Guide to Animal Protection*, 181, "Like sheep, goats have been used in military research as living targets for studying the effects of gunshot wounds on the battlefield. They have also had nuclear bombs detonated near them in order for researchers to study the effects of nuclear explosions and radiation from fallout."
32 Murugan and Raman, *Story of a Goat*, 179.
33 Kumar, "How Perumal Murugan Was Resurrected."
34 Polymerase chain reaction, or PCR, is a laboratory technique used to make multiple copies of a segment of DNA.
35 I feel deep empathy for any animal who experiences mastitis. As a new mother to Grace, I had dried and cracked nipples and developed mastitis. It was extremely painful, and I passed out one day because of a high fever. Even though I tried to avoid taking antibiotics for fear of exposing Grace to medication, I was treated with them, and the infection cleared up in two days.
36 Thanks again to Darcy Gervasio. *OED Online*, s.v. "throwing," accessed June 11, 2020, www.oed.com.
37 A *dope* is the substance remaining in a vial from wet-spinning spider silk protein dissolved in solvents to make a polymer. Spinning dopes is a common practice in the lab to make the silk protein usable.
38 Remarkably, given the lure of biotech as a path to outrageous financial gain, scientists opt to maintain a grant-funded-university lifestyle. Finally, at the end of several days of interviews, I simply came out and asked Justin, "Do you ever think, like, you're going to get rich doing this?" He chuckled a bit. "Never," he said. "It's never been my goal, right? No, I know, I know, you think about it. It's an interesting question. Because we have people pinging us all the time over different applications. And this is just a good example: You have a major tire manufacturer for a bicycle tire, and they want to innovate spider silk for the tires. And you know, they're going to approach you to put, you know, your product, whatever that may be, into a bicycle tire. I'll never be able to afford that bike. Or you can start working on things like sutures or retina replacement and things that will benefit everybody."
39 DeMello, *Animals and Society*.
40 Morris, *Euthanasia in Veterinary Medicine*.
41 Taylor, "'Never an It.'"
42 Wilkie, *Livestock/Deadstock*.
43 See US Department of Agriculture, and Animal and Plant Health Inspection Service, "Animal Welfare Act and Animal Welfare Regulations."
44 US Food and Drug Administration, "Animals with Intentional Genomic Alterations."

45 The FDA's guidance on regulation of genetically engineered (GE) animals recommends a review that includes seven categories:
    - *Product definition:* a broad statement characterizing the intentional genomic alteration (IGA) in the GE animal and the claim being made for it
    - *Molecular characterization of the construct (the IGA):* a description of the rDNA construct or other genomic alteration and how it is produced
    - *Molecular characterization of the GE animal lineage:* a description of the method by which the rDNA construct or other genomic alteration was introduced into the animal and whether it is stably maintained over time
    - *Phenotypic characterization of the GE animal:* comprehensive data on the characteristics of the GE animal and its health
    - *Durability plan:* the sponsor's plan to demonstrate that the alteration will remain the same over time, and continue to have the same effect
    - *Environmental impact and food/feed safety:* the assessment of any environmental impacts and, for GE animals of food species, an assessment of the safety of food derived from those GE animals showing it is safe to eat for humans and/or animals
    - *Claim validation:* a demonstration that the animal containing the rDNA construct/IGA has the characteristics that the developer says it has
46 US Food and Drug Administration, "Guidance for Industry Regulation."
47 Star and Griesemer, "Institutional Ecology, 'Translations' and Boundary Objects."
48 Gibson, *Ecological Approach to Visual Perception.*
49 Norman, *Design of Everyday Things.*
50 Special thanks to Darcy Gervasio. *OED Online,* s.v. "cull," accessed June 11, 2020, www.oed.com.
51 Mehrabi, "Making Death Matter," 234.

3. THE GOAT AS A SYSTEM

1 *Donor insemination* is the preferred term among many people.
2 On historical alternatives to heterosexual intercourse for reproduction, see Ombelet and Van Robays, "Artificial Insemination History." Regarding the stigma associated with technology-assisted reproduction, see, for example, Faccio, Iudici, and Cipolletta, "To Tell or Not to Tell?"; Kleinert et al., "Motives and Decisions for and against Having Children."
3 This reminds me of a painful encounter I had when letting other parents know that my girlfriend and I were splitting up. One parent pulled me aside and said, "This is so sad. And sad for Charlotte too, since we always talked about you as one of the examples of a lesbian family that was working." I think she thought she was empathizing.
4 A technique called *premium wash* is described in Fertility Center of California, "Sperm Washing: Sperm Washing and Preparation Techniques for Artificial Insemination and IVF," accessed September 15, 2021, www.spermbankcalifornia.

com: "This method uses density gradient centrifugation to isolate and purify the motile sperm in order to obtain a sperm sample with a motility of at least 90%, depending on the initial quality of the sample. Different concentrations of isolate (extremely dense fluid) are layered in a test tube in an ascending order of density (heaviest layer at the bottom). When a semen sample is placed upon the upper-most isolate layer and centrifuged, any debris, round cells, non-motile and poor quality sperm remain in the top layers. Only the motile sperm are able to get through to the bottom layer and are then concentrated for use in artificial insemination. This procedure takes 1 hour. The premium sperm wash technique is excellent for fresh or frozen sperm and can help assess male factor fertility."

5   Douglas, *Purity and Danger*.
6   Stewart, "Politics of Cultural Theory."
7   Darwin, *Origin of Species*; Darwin, *Voyage of the Beagle*; Heiligman, *Charles and Emma*; Browne, *Charles Darwin: Voyaging*, 1; Browne, *Power of Place*.
8   Darwin, "Darwin Correspondence Project."
9   See Stengers, "Autonomy and the Intrusion of Gaia."
10 Williams, *Country and the City*.
11 Fish, *Living Factories*.
12 Bennett, *Enchantment of Modern Life*.
13 Kljajevic et al., "Seasonal Variations of Saanen Goat Milk Composition.
14 As Bri Bell explained, "The volume we run depends on the protein we are processing. MaSp1 we run through about ten liters at a time, while the MaSp2 we process twenty to forty liters at a time. The difference is from the difference in protein concentration in the milk, with MaSp2 typically being around one-tenth to two-tenth grams per liter, while MaSp1 has traditionally been around one or two grams per liter. Processes are for twenty hours unless we are doing a very small run of five liters, which then would only run for five hours."
15 Latour, *Science in Action*.
16 Bri Bell explained the multiple factors that have gone into refinements and redesigns of the purification protocols: "We have gone through several process modifications for both protein types. Changes that are incorporated into both protein purification methods are:
    - "running the process for 20 hours instead of 48 hours
    - "optimizing the ammonium sulfate needed for precipitating the spider silk proteins
    - "running the milk and arginine solution near the protein isoelectric point
    - "while not a process modification, both proteins are run under the "supervision" of level controls to prevent draining out the milk solution or spider silk solution. This process has changed and improved over time as we learn more about available sensors and flow control
    - "the addition of PMSF [phenylmethylsulphonyl fluoride] to decrease the effects of protease activity on the spider silk protein
    - "process at room temperature instead of in a cold room (4C)"

17 See, for example, Levine, "Symbolism of Milk and Honey"; Bramwell, "Blood and Milk."
18 Legal scholar Andrea Freeman explores food oppression against socially marginalized groups. For example, milk and other dairy products, not easily digested by Americans with non-European backgrounds, have been heavily marketed and subsidized to become a staple of the USDA's food education practices. Freeman, "Unbearable Whiteness of Milk."
19 DuPuis, *Nature's Perfect Food*.
20 Several food policy papers evaluate the economic impact of food labeling on purchases and discuss the positive associations that consumers have for terms associated with purity. See, for example, McFadden and Huffman, "Willingness-to-Pay."
21 Almost everything about the milk we purchase in the grocery store is processed. See Arumugam, "Milk Isn't As 'Natural' As You Think."
22 Bri's developing sensitivities to alternative ways of perceiving influences on the milk run brings to mind Keller, *A for the Organism*.
23 My colleague Stephen Cooke, a biochemist, clarified for me the use of filtration for proteins: "The column is simply a tube packed with small spherical beads. The size of these beads will limit what can pass through. If the beads are 30–50 kilodaltons each, then they will allow molecules lighter than this through easily, but for something bigger, like the 65 kilodalton silk protein, it will get stuck on the column and, hence, is separated from most of the milk. In reality it is likely that the 65 kilodalton protein can be forced through the column by applying pressure but it will take *much* longer to pass through than all of the other constituents.

    "So, as a simplified summary, the column is a sieve, and the 30–50 kilodaltons are a measure of the size of the holes in the sieve. The 65 kilodalton protein is too large to go through the holes in the sieve but almost everything else will and so separation is achieved."
24 Stephen Cooke also provided me with this handy guidelines on molecular sizes in units of daltons (Da), from smallest to largest:
    One hydrogen molecule: 2 Da
    Atmospheric chemicals (air, water, carbon dioxide, etc.): ~10–50 Da
    Amino acids, common sugars, and common pharmaceuticals: ~100–300 Da
    Fats: ~100–500 Da
    Steroids: ~300 Da
    Many catalysts: ~800 Da
    Nanoparticles: ~1,000 Da and up
    Peptides: ~1,000–5,000 Da
    Proteins: ~5,000 Da and up
    Viruses: ~1,000,000 Da and up
25 Rutherford, "Synthetic Biology," 75–84.
26 Rutherford, "Synthetic Biology," 76.
27 Rutherford, *Creation*, 150.

28 Rutherford, *Creation*, 148.
29 Davies, *Synthetic Biology*.
30 See, for example, Shapiro, *How to Clone a Mammoth*.
31 Roosth, *Synthetic*, 15.
32 Roosth, *Synthetic*, 55.
33 Croce, Grilli, and Murtinu, "Venture Capital Enters Academia."
34 Edwards and Roy, "Academic Research in the 21st Century."
35 For an excellent analysis of how ownership is managed in academic labs, see Johnson, *Commercialism and Conflict in Academic Science*.
36 Moore, *Sperm Counts*.
37 Shapin, *Scientific Life*.
38 Mirowski, *Science-Mart: Privatizing American Science*.
39 Donna Haraway has certainly trodden this territory before as she investigated the creation of OncoMouse (circa 1988), a living organism invented at Harvard University and patented. Ownership then transferred from Harvard to DuPont as corporate biology created a new world order for our kin. Spider goats are adjacent to OncoMouse, emerging two years later, though the goats themselves are not the medical device or product; rather, they are a system to produce a product. See Haraway, *Modest_Witness@Second_Millennium.FemaleMan©_Meets_OncoMouse™*.
40 Morgan, *Laboring Women*.
41 Baldwin, *Synthetic Biology: A Primer*.
42 Perlo, "Marxism and the Underdog."
43 Noske, *Beyond Boundaries*.
44 Dickens, *Reconstructing*.
45 Haraway, *Companion Species Manifesto*.
46 Shaka McGlotten's work on Black data is very instructive here, as they argue that bodies that rank lower in systems of stratification (brown, female, queer) are the producers of raw materials used for data-driven economies. Nonhuman animals are also producers, rather than consumers. See McGlotten, "Streaking."
47 Haraway, "Cyborg Manifesto."
48 Isabelle Stengers suggested the idea of a disposition to invention.
49 Latour, *Pasteurization of France*.

4. THIN SKINNED

1 Kevlar, designed by the DuPont Company and commercially available in the 1970s, is described as a fiber made of superstrong, rigid polymer molecules belonging to a small class called aramids.
2 An exemplar from the news Santhanam, "Gun Deaths Started to Rise."
3 Companies offer different degrees of coverage from ballistic threat in emergency kits. For example, at Ready-to-go Survival (https://readytogosurvival.com), you can purchase ballistic shields for backpacks, and BulletBlocker (www.bulletblocker.com) sells the School Safety and Survival pack for $450 and offers 160 square inches of ballistic coverage.

4 In this chapter I specifically use the term *products* to refer to objects as commodities, material things that can be made and possibly sold. Clearly, there are other ways of thinking about products. A whole slew of educational achievements and degrees were funded; many herds of goats were made; life (of goats and spiders) was created and ended; and knowledge, machines, and networks of collaborating people were produced. All these developments are connected to the sociology of the spider goat. I'm thinking here of Althusser's aleatory materialisms and Lucretius's *On the Nature of Things*.
5 I was trained to do this mapping many years ago by my dissertation director, Adele Clarke. See for example, Clarke, *Situational Analysis*.
6 This brings to mind the late 2010s meme of "starter packs," the curation of photos, gifs, texts, and screenshots collected to represent an identity, subculture, or cohort. For the urban young upper-middle-class white mom, it might be a Lululemon logo, a large skim oat milk latte in a cup from a fancy café, a looping gif of a woman drinking a large glass of white wine, and a list of gender-neutral baby names. The accuracy of these starter packs can be surprising—as well the reliance on brands associated with that demographic. What does it mean if you feel you need the objects in the starter pack that most closely describes you? Whose starter pack includes spider silk products?
7 The list of these companies includes Bolt Threads, Kraig Biocraft Laboratories, Spiber, and AM Silk.
8 Sigmund Freud offered up five elements of anxiety: fear of uncertainty (unknown), fear of abandonment (loss of love), fear of bodily harm or damage (castration), fear of punishment (castration, loss of love, guilt), and morality (fear of one's own superego, guilt). Donald Winnicott suggests we have *primitive agonies*, Melanie Klein coined *persecutory anxiety*, and Wilfred Bion explained the *nameless dread*. No doubt other psychoanalysts proposed similar pre-oedipal or oedipal stages of developmental, foundational fear and anxiety. This is the child's preverbal experience of the world when certain traumas may create lifelong (functional or dysfunctional) mechanisms for coping (or not) with everyday life.
9 Winnicott, *Child, the Family, and the Outside World*.
10 For "universal phenomenon," see Ogden, "Fear of Breakdown." The unit itself is "a state in which the infant is a unit, a whole person, with an inside and an outside, and a person living in a body, and more or less bounded by the *skin*" (Winnicott, Shepherd, and Davis, "Fear of Breakdown," 87–95 [emphasis mine]).
11 As my friend Tine Pahl, a psychoanalyst, has noted in our conversations, "Maybe this early primitive agony is also the reason we're all so utterly obsessed with superheroes."
12 Paul Siegel, email correspondence with author, September 20, 2018.
13 Winnicott, Shepherd, and Davis, "Fear of Breakdown," 90.
14 Ogden, "Fear of Breakdown," 213.
15 A recent modeling economic study of spider silk production using *E. coli* determined that the cost of production is $23 per kilogram of spider silk protein. We

can surmise that it would be more costly using goats. Edlund et al., "Synthetic Spider Silk Production from *Escherichia coli*."
16 This is a good reminder of Raymond Williams's essay on how discoveries are often accidents. See Williams, "Technology and the Society (1974)."
17 This sparks the question of how much spider silk protein is needed for an application. As described earlier, a good yield is considered one gram per liter of goat milk. Justin explained (emphasis mine), "The quantity of silk from *non-silkworm* sources entirely depends on the application. For example, a catheter coating would likely only require a couple of milligrams (if that) to coat the entire catheter. A garment on the other hand requires considerably more . . . say, 500 [grams] per garment. Dental applications would require in the single-gram quantities per pre-loaded syringe. That syringe would then be used for multiple sites in the oral cavity and used on multiple patients (potentially).

"As you can quickly surmise, medical applications are low volume and high value, which is much more suitable for current spider silk production levels. However, the goats, *E. coli* and alfalfa (as well as yeast) are not well suited to the garment or textile industries. The fiber qualities are just not good enough coming out of these systems. Silkworms are a much better system to produce fibers as they spin the silk for us using cellular machinery that is mimetic to the specialized machinery that spiders use to spin their silk. The economics of producing spider silk in this fashion will be much more economical than recombinant production from other systems as an entire industry exists to produce silkworm silk (and one that is hungry for improved silk)."
18 Utah State University, Technology Transfer Services, "Periodontal Disease Treatment with Spider Silk Gel," accessed September 8, 2021, https://research.usu.edu.
19 Transvaginal meshes, most commonly made of polypropylene plastic, are used to treat stress incontinence and pelvic organ prolapse, a condition where the bladder, uterus, and other organs sag into the vagina. A synthetic mesh has been used since the 1996 for stress incontinence, and the FDA approved it in 2002 for pelvic organ prolapse through a "510k clearance" loophole that enabled the device to be cleared through its similarity to other devices without testing on women's bodies. Organ perforation, vaginal wall erosion, and severe pain are some of the complications. There have been tens of thousands of lawsuits against manufacturers of vaginal mesh. In 2019, the FDA banned the sale of mesh for pelvic organ prolapse repair.
20 Roosth, *Synthetic*.
21 For an exemplar article expressing the anxiety of the textile industry's reliance on synthetic fibers, see Van Haren, "Alternatives to Petro-Chemical-Based Materials."
22 Zhang et al., "Nitrogen Inaccessibility Protects Spider Silk."
23 See David Breslauer, "Co-founder and Chief Science Officer at Bolt Threads," LinkedIn profile page, www.linkedin.com.
24 See, for example, Museum of Modern Art, *Items: Is Fashion Modern?*, October 1, 2017, www.moma.org.

25 The company's website describes where the product's name comes from: "Our key ingredient, b-silk™ protein, is a high molecular weight protein made up of a series of 18 repeating molecular segments (we call them blocks, which is where the 'b' comes from). Each of these blocks is made up of an amino acid pattern that was derived from the pattern found in natural silk. We drew inspiration from its structure and created Eighteen B" (Bolt Threads, "Meet the Beebe Lab: Beta Testing at Its Most Beautiful," www.eighteenb.com). The FAQ section from Beebe Labs states, "No spiders are harmed in the making of b-silk protein! In fact, there are no spiders at all in the process. We originally studied real spider silk to understand the relationship between the spider DNA and the characteristics of the fibers they make. Today's technology allows us to make those proteins without using spiders, which is a big relief to the arachnophobes among our team" (Bolt Threads, "Bolt Technology: Meet B-Silk Protein," https://boltthreads.com).

26 Even though there is less regulation of US beauty products than there is in the European Union, the industry itself sees benefits to making and marketing "clean" products. The clean beauty market describes the use of nontoxic, environmentally sustainable, organic, or vegan ingredients. However, no agency is responsible for regulating what is considered clean. The use of the term *clean* in marketing bears similarities to greenwashing. See Fleming and Rosenstein, "Ultimate Guide to Clean Beauty," for an example of how the popular press operationalizes and celebrates clean.

27 The company website https://boltthreads.com promotes and explains its production and use of spider silk protein.

28 I've attempted situational analysis as outlined by Clarke, Friese, and Washburn, *Situational Analysis: Grounded Theory after the Interpretive Turn*, and assemblage as described by Deleuze and Guattari, *A Thousand Plateaus*.

29 Berlant, *Cruel Optimism*; Tsing, *Mushroom at the End of the World*.

30 Berlant, *Cruel Optimism*, 1.

31 As I frequently told my daughter when she was ten years old, you don't need a cell phone; you want one. This is my third time in this argument, so I might be a pro by now. But then when she turned eleven and walked home from school solo, she declared she needed one (a need that I realize is also bullshit).

CONCLUSION

1 Haraway, *Staying with the Trouble*.
2 Nagel, "What Is It Like to Be a Bat?"
3 In being anti-reductionist, Nagel is claiming that subjective experiences are not reducible to objective characteristics. Some facts require a first-person (animal) perspective and cannot be knowable by humans.
4 Quammen, "Virus, the Bats and Us."
5 By "war machines," I include the whole apparatus of military infrastructure, technology, blood capital, and geopolitics, as distinct from Deleuze's notion.

6   I also want readers to be as compelled by the spiders (the forgotten species long ago sequenced) as they are by the goats. But this desire could potentially compromise my "data," because I tell the story to be most generous to my various informants, possibly corrupting a stance of detachment valued by some scholars.
7   Cussins, "Ontological Choreography."
8   See work by Thom van Dooren on mourning, in particular Dooren, *Flight Ways*; Dooren, "Mourning Crows."
9   On the types of human mourning for nonhuman animals, see, for example, Buchanan, "Bear Down."
10 Yusoff, "Aesthetics of Loss."
11 Atwood, *Year of the Flood*.
12 This number is not static, as new dogs breeds are added periodically. See, for example, Asmelash, "American Kennel Club Announced Two New Breeds."
13 See "Ethics of Raising Purebred Dogs."
14 Mel Chen, *Animacies*.

REFERENCES

Adams, Carol J. *The Sexual Politics of Meat: A Feminist-Vegetarian Critical Theory*. New York: Continuum, 1995.
Allport, Gordon W. *The Nature of Prejudice*. Unabridged, 25th anniv. ed. Reading, MA: Addison-Wesley, 1979.
Amundson, Carol A. *How to Raise Goats: Everything You Need to Know; Meat, Milk, Fiber & Pet Goats, Breed Guide & Purchasing, Proper Care & Healthy Feeding, Showing Advice*. Minneapolis: Voyageur Press, 2009.
Anthes, Emily. *Frankenstein's Cat: Cuddling Up to Biotech's Brave New Beasts*. New York: Scientific American / Farrar, Straus and Giroux, 2013.
Arluke, Arnold, and Clinton Sanders. *Regarding Animals*. Philadelphia: Temple University Press, 1996.
Arumugam, Nadia. "Why Milk Isn't As 'Natural' As You Think." *Forbes*, December 3, 2012. www.forbes.com.
Asmelash, Leah. "The American Kennel Club Announced Two New Breeds: The Barbet and Dogo Argentino." *CNN*, January 3, 2020. www.cnn.com.
Atkinson, Michael. *Tattooed: The Sociogenesis of a Body Art*. Toronto: University of Toronto Press, 2003.
Atwood, Margaret. *The Year of the Flood*. New York: Anchor, 2009.
Baldwin, Geoff, Travis Bayer, Robert Dickinson, Tom Ellis, Paul S. Freemont, Richard I. Kitney, Karen Polizzi, and Guy-Bart Stan, eds. *Synthetic Biology: A Primer*, rev. ed. London: World Scientific Publishing Co. Pte. Ltd., 2016.
Banister, Elizabeth M. "Evolving Reflexivity: Negotiating Meaning of Women's Midlife Experience." *Qualitative Inquiry* 5, no. 1 (March 1999): 3–23. https://doi.org/10.1177/107780049900500101.
Barad, Karen. *Meeting the Universe Halfway: Quantum Physics and the Entanglement of Matter*. Durham, NC: Duke University Press, 2007.
Beck, Ulrich. *Risk Society: Towards a New Modernity*. Theory, Culture & Society. Newbury Park, CA: SAGE Publications, 1992.
Benjamin, Ruha. *People's Science: Bodies and Rights on the Stem Cell Frontier*. Stanford, CA: Stanford University Press, 2013.
Bennett, Jane. *The Enchantment of Modern Life: Attachments, Crossings, and Ethics*. Princeton, NJ: Princeton University Press, 2001.
Berlant, Lauren Gail. *Cruel Optimism*. Durham, NC: Duke University Press, 2011.

Billo, Emily, and Nancy Hiemstra. "Mediating Messiness: Expanding Ideas of Flexibility, Reflexivity, and Embodiment in Fieldwork." *Gender, Place & Culture* 20, no. 3 (May 2013): 313–328. https://doi.org/10.1080/0966369X.2012.674929.

Bishop, Jerry. "Army Genetic Scientist Says He's Found a Way to Mass-Produce Tough Spider Silk." *Wall Street Journal*, February 26, 1990.

Blanchette, Alex. *Porkopolis: American Animality, Standardized Life, and the Factory Farm.* Durham, NC: Duke University Press, 2020.

Bloch, Natalia. "Making a Community Embedded in Mobility." *Transfers* 8, no. 3 (December 1, 2018): 36–54. https://doi.org/10.3167/TRANS.2018.080304.

Bordo, Susan. *Unbearable Weight: Feminism, Western Culture, and the Body.* 10th anniv. ed., Berkeley, CA: University of California Press, 2013.

Bramwell, Ros. "Blood and Milk: Constructions of Female Bodily Fluids in Western Society." *Women & Health* 34, no. 4 (December 13, 2001): 85–96. https://doi.org/10.1300/J013v34n04_06.

Browne, E. Janet. *The Power of Place.* Vol. 2 of *Charles Darwin, a Biography.* Princeton, NJ: Princeton University Press, 2002.

———. *Voyaging.* Vol. 1 of *Charles Darwin, a Biography.* Princeton, NJ: Princeton University Press, 1996.

Buchanan, Brett. "Bear Down: Resilience and Multispecies Ethology." In *The Routledge Companion to the Environmental Humanities*, edited by Ursula K. Heise, Jon Christensen, and Michelle Niemann, 289–298. London: Routledge, Taylor & Francis Group, 2016.

Burke, Sara E., John F. Dovidio, Julia M. Przedworski, Rachel R. Hardeman, Sylvia P. Perry, Sean M. Phelan, David B. Nelson, Diana J. Burgess, Mark W. Yeazel, and Michelle van Ryn. "Do Contact and Empathy Mitigate Bias against Gay and Lesbian People among Heterosexual First-Year Medical Students? A Report from the Medical Student CHANGE Study." *Academic Medicine* 90, no. 5 (May 2015): 645–551. https://doi.org/10.1097/ACM.0000000000000661.

Candea, Matei. "'I Fell in Love with Carlos the Meerkat': Engagement and Detachment in Human Animal Relations." *American Ethnologist* 37, no. 2 (2010): 241–258. https://doi.org/10.1111/j.1548-1425.2010.01253.x

Carbone, Larry. "Pain in Laboratory Animals: The Ethical and Regulatory Imperatives." *PLoS ONE* 6, no. 9 (September 7, 2011): e21578. https://doi.org/10.1371/journal.pone.0021578.

Chen, Mel. *Animacies: Biopolitics, Racial Mattering, and Queer Affect.* Durham, NC: Duke University Press, 2012.

Choudhary, Srishti. "Pastoralists of Himachal Pradesh an Unusually Casualty of Global Warming." *LiveMint*, June 26, 2019. www.livemint.com.

Chris, Cynthia. *Watching Wildlife.* Minneapolis: University of Minnesota Press, 2006.

Clark, J. L. "Killing the Enviropigs." *Journal of Animal Ethics* 5, no. 1 (2015): 20. https://doi.org/10.5406/janimalethics.5.1.0020.

Clarke, Adele E. *Situational Analysis: Grounded Theory after the Postmodern Turn.* Thousand Oaks, CA: SAGE Publications, 2005.

Clarke, Adele E., Carrie Friese, and Rachel Washburn, *Situational Analysis: Grounded Theory after the Interpretive Turn*, 2nd ed. (Los Angeles: SAGE Publications, 2018).

Clarke, Adele E., and Joan H. Fujimura, eds. *The Right Tools for the Job: At Work in 20th Century Life Sciences*. Princeton, NJ: Princeton University Press, 2014.

Cohen, Mathilde. "Toward an Interspecies Right to Breastfeed." *Animal Law Review* 26 (March 21, 2019): 1–40. https://ssrn.com/abstract=3581590

Crandall, Emily. "Interview: Zakiyyah Iman Jackson on Becoming Human." *Always Already Podcast*. Accessed July 1, 2021. https://alwaysalreadypodcast.wordpress.com.

Croce, Annalisa, Luca Grilli, and Samuele Murtinu. "Venture Capital Enters Academia: An Analysis of University-Managed Funds." *Journal of Technology Transfer* 39, no. 5 (October 2014): 688–715. https://doi.org/10.1007/s10961-013-9317-8.

Cussins, Charis. "Ontological Choreography: Agency through Objectification in Infertility Clinics." *Social Studies of Science* 26, no. 3 (August 1996): 575–610. https://doi.org/10.1177/030631296026003004.

Dalton, Stephen. *Spiders: The Ultimate Predators*. Buffalo, NY: Firefly Books, 2011.

Darwin, Charles. "Darwin Correspondence Project." *Letter No. 468, To Emma Wedgwood [20 January 1839]*. University of Cambridge. Accessed July 2, 2021. www.darwinproject.ac.uk.

———. *On the Origin of Species*. Washington Square, NY: New York University Press 1988; original 1859.

———. *The Voyage of the Beagle: The Illustrated Edition of Charles Darwin's Travel Memoir and Field Journal*. Minneapolis: Zenith Press, 2015.

D'Aulaire, Ingri, and Edgar Parin d'Aulaire. *Ingri and Edgar Parin d'Aulaire's Book of Greek Myths*. Garden City, NY: Doubleday & Company, 1962.

Davids, Tine, and Karin Willemse. "Embodied Engagements: Feminist Ethnography at the Crossing of Knowledge Production and Representation—An Introduction." *Women's Studies International Forum* 43 (March 2014): 1–4. https://doi.org/10.1016/j.wsif.2014.02.001.

Davies, Jamie A. *Synthetic Biology: A Very Short Introduction*. Very Short Introductions; 573. Oxford: Oxford University Press, 2018.

DellaPosta, Daniel. "Gay Acquaintanceship and Attitudes toward Homosexuality: A Conservative Test." *Socius: Sociological Research for a Dynamic World* 4 (January 2018): 1–12. https://doi.org/10.1177/2378023118798959.

Deleuze, Gilles, and Félix Guattari. *A Thousand Plateaus: Capitalism and Schizophrenia*. Minneapolis: University of Minnesota Press, 1987.

DeMello, Margo. *Animals and Society: An Introduction to Human-Animal Studies*. New York: Columbia University Press, 2012.

Dickens, Peter. *Reconstructing Nature: Alienation, Emancipation, and the Division of Labour*. London; New York: Routledge, 1996.

Dooren, Thom van. *Flight Ways: Life and Loss at the Edge of Extinction*. New York: Columbia University Press, 2014.

———. "Mourning Crows: Grief and Extinction in a Shared World." In *Routledge Handbook of Human-Animal Studies*, edited by Garry Marvin and Susan McHugh, 275–289. London: Routledge, Taylor & Francis Group, 2014.

Douglas, Mary. *Purity and Danger: An Analysis of the Concepts of Pollution and Taboo*. London: Routledge, 1978.

Dowden, Ken. *Zeus. Gods and Heroes of the Ancient World*. London; New York: Routledge, 2006.

DuPuis, E. Melanie. *Nature's Perfect Food: How Milk Became America's Drink*. New York: New York University Press, 2002.

Edlund, Alan M., Justin Jones, Randolph Lewis, and Jason C. Quinn. "Economic Feasibility and Environmental Impact of Synthetic Spider Silk Production from *Escherichia coli*." *New Biotechnology* 42 (May 2018): 12–18. https://doi.org/10.1016/j.nbt.2017.12.006.

Edwards, Marc A., and Siddhartha Roy. "Academic Research in the 21st Century: Maintaining Scientific Integrity in a Climate of Perverse Incentives and Hypercompetition." *Environmental Engineering Science* 34, no. 1 (January 2017): 51–61. https://doi.org/10.1089/ees.2016.0223.

Ekberg, Merryn. "The Parameters of the Risk Society: A Review and Exploration." *Current Sociology* 55, no. 3 (May 2007): 343–366. https://doi.org/10.1177/0011392107076080.

"The Ethics of Raising Purebred Dogs." Editorial. *New York Times*, February 12, 2013. www.nytimes.com.

Faccio, Elena, Antonio Iudici, and Sabrina Cipolletta. "To Tell or Not to Tell? Parents' Reluctance to Talking About Conceiving Their Children Using Medically Assisted Reproduction." *Sexuality & Culture* 23, no. 2 (June 2019): 525–543. https://doi.org/10.1007/s12119-019-09586-7.

Fish, Kenneth. *Living Factories: Biotechnology and the Unique Nature of Capitalism*. Montréal: McGill-Queen's University Press, 2013.

Fleming, Olivia, and Jenna Rosenstein. "The Ultimate Guide to Clean Beauty." *Harper's Bazaar*, April 22, 2020. www.harpersbazaar.com.

Franklin, Sarah. *Dolly Mixtures: The Remaking of Genealogy*. Durham, NC: Duke University Press, 2007.

Freeman, Andrea. "The Unbearable Whiteness of Milk: Food Oppression and the USDA." *UC Irvine Law Review* 3 (2013): 1251–1280. https://ssrn.com/abstract=2338749.

Gibson, James J. *The Ecological Approach to Visual Perception*. New York; London: Psychology Press, 2015.

Giddens, Anthony. *Modernity and Self-Identity: Self and Society in the Late Modern Age*. Cambridge, UK: Polity Press, 1991.

———. *The Third Way: The Renewal of Social Democracy*. Malden, MA: Polity Press, 1999.

Giddens, Anthony, and Christopher Pierson. *Conversations with Anthony Giddens: Making Sense of Modernity*. Stanford, CA: Stanford University Press, 1998.

Goldstein, Jesse, and Elizabeth Johnson. "Biomimicry: New Natures, New Enclosures." *Theory, Culture & Society* 32, no. 1 (January 2015): 61–81. https://doi.org/10.1177/0263276414551032.

Grazian, David. *American Zoo: A Sociological Safari*. Princeton, NJ: Princeton University Press, 2018. https://doi.org/10.1515/9781400873616.

Gruen, Lori. *Entangled Empathy: An Alternative Ethic for Our Relationships with Animals*. New York: Lantern Books, 2015.

Gruen, Lori, and Kari Weil. "Animal Others—Editors' Introduction." *Hypatia* 27, no. 3 (2012): 477–487. https://doi.org/10.1111/j.1527-2001.2012.01296.x.

Hanley, Sean. "The Whelming Sea." Master's thesis, CUNY, 2020. https://academicworks.cuny.edu.

Haraway, Donna. "A Cyborg Manifesto: Science, Technology, and Socialist-Feminism in the Late Twentieth Century." In *Simians, Cyborgs and Women: The Reinvention of Nature*. New York: Routledge, 1991.

———. *Companion Species Manifesto: Dogs, People, and Significant Otherness*. Chicago: Prickly Paradigm Press, 2003.

———. *Modest_Witness@Second_Millennium.FemaleMan©_Meets_OncoMouse™*. New York: Routledge, 1997.

———. *Staying with the Trouble: Making Kin in the Chthulucene*. Experimental Futures: Technological Lives, Scientific Arts, Anthropological Voices. Durham, NC: Duke University Press, 2016.

Haraway, Donna, and Thyrza Nichols Goodeve. *How like a Leaf: An Interview with Thyrza Nichols Goodeve*. New York: Routledge, 2000.

Hartigan, John. *Shaving the Beasts: Wild Horses and Ritual in Spain*. Minneapolis: University of Minnesota Press, 2020.

Heiligman, Deborah. *Charles and Emma: The Darwins' Leap of Faith*. New York: Henry Holt and Co., 2009.

Hekmat, Sharareh, and Lindsay Nicole Dawson. "Students' Knowledge and Attitudes towards GMOs and Nanotechnology." *Nutrition & Food Science* 49, no. 4 (July 8, 2019): 628–638. https://doi.org/10.1108/NFS-07-2018-0193.

Hennessey, Rachel. "GMO Food Debate in the National Spotlight." *Forbes*, November 3, 2012. www.forbes.com.

Herek, Gregory M., and Eric K. Glunt. "Interpersonal Contact and Heterosexuals' Attitudes toward Gay Men: Results from a National Survey." *Journal of Sex Research* 30, no. 3 (August 1993): 239–244. https://doi.org/10.1080/00224499309551707.

High, Kathy. "Playing with Rats." In *Tactical Biopolitics: Art, Activism and Technoscience*, edited by Beatriz Da Costa and Kavita Philip. Cambridge, MA: MIT Press, 2008.

Hill Collins, Patricia. "Learning from the Outsider Within: The Sociological Significance of Black Feminist Thought." *Social Problems* 33, no. 6 (1986): 14–32. https://doi.org/10.2307/800672.

Hochschild, Arlie Russell, and Anne Machung. *The Second Shift: Working Families and the Revolution at Home*. New York: Penguin Books, 2012.

Hoffman, Yaakov S. G., Shani Pitcho-Prelorentzos, Lia Ring, and Menachem Ben-Ezra. "'Spidey Can': Preliminary Evidence Showing Arachnophobia Symptom Reduction Due to Superhero Movie Exposure." *Frontiers in Psychiatry* 10 (June 7, 2019): 354. https://doi.org/10.3389/fpsyt.2019.00354.

Ingold, Tim. "When ANT Meets SPIDER: Social Theory for Arthropods." In *Material Agency*, edited by Carl Knappett and Lambros Malafouris, 209–215. Boston: Springer US, 2008. https://doi.org/10.1007/978-0-387-74711-8_11.

Jackson, Zakiyyah Iman. *Becoming Human: Matter and Meaning in an Antiblack World*. Sexual Cultures. New York: New York University Press, 2020.

Johnson, David R. *A Fractured Profession: Commercialism and Conflict in Academic Science*. Baltimore: Johns Hopkins University Press, 2017.

Joseph, Suad. "Relationality and Ethnographic Subjectivity: Key Informants and the Construction of Personhood in Fieldwork." In *Feminist Dilemmas in Fieldwork*, edited by Diane L. Wolf, 107–121. New York: Routledge, 2018.

Joshi, M. B. "Phylogeography and Origin of Indian Domestic Goats." *Molecular Biology and Evolution* 21, no. 3 (December 5, 2003): 454–462. https://doi.org/10.1093/molbev/msh038.

Kalof, Linda, Joe Zammit-Lucia, Jessica Bell, and Gina Granter. "Fostering Kinship with Animals: Animal Portraiture in Humane Education." *Environmental Education Research* 22, no. 2 (February 17, 2016): 203–228. https://doi.org/10.1080/13504622.2014.999226.

Keller, Evelyn Fox. *A Feeling for the Organism: The Life and Work of Barbara McClintock*. San Francisco: W. H. Freeman, 1983.

King, Barbara J. *How Animals Grieve*. Chicago: University of Chicago Press, 2014.

Kleinert, Evelyn, Olaf Martin, Elmar Brähler, and Yve Stöbel-Richter. "Motives and Decisions for and against Having Children among Nonheterosexuals and the Impact of Experiences of Discrimination, Internalized Stigma, and Social Acceptance." *Journal of Sex Research* 52, no. 2 (February 12, 2015): 174–185. https://doi.org/10.1080/00224499.2013.838745.

Kljajevic, Nemanja V., Igor B. Tomasevic, Zorana N. Miloradovic, Aleksandar Nedeljkovic, Jelena B. Miocinovic, and Snezana T. Jovanovic. "Seasonal Variations of Saanen Goat Milk Composition and the Impact of Climatic Conditions." *Journal of Food Science and Technology* 55, no. 1 (January 2018): 299–303. https://doi.org/10.1007/s13197-017-2938-4.

Kumar, Amitava. "How Perumal Murugan Was Resurrected through Writing." *New Yorker*, December 12, 2019. www.newyorker.com.

Kuntner, Matjaž, Chris A. Hamilton, Ren-Chung Cheng, Matjaž Gregorič, Nik Lupše, Tjaša Lokovšek, Emily Moriarty Lemmon, Alan R. Lemmon, Ingi Agnarsson, Jonathan A. Coddington, and Jason E. Bond. "Golden Orbweavers Ignore Biological Rules: Phylogenomic and Comparative Analyses Unravel a Complex Evolution of Sexual Size Dimorphism." *Systematic Biology* 68, no. 4 (December 4, 2018): 555–572. https://doi.org/10.1093/sysbio/syy082.

Latour, Bruno. *Science in Action: How to Follow Scientists and Engineers through Society*. 1987. Reprint, Cambridge, MA: Harvard University Press, 2003.
———. *The Pasteurization of France*. First Harvard University Press paperback ed. Cambridge, MA: Harvard University Press, 1993.
Laurent-Simpson, Andrea, and Celia C. Lo. "Risk Society Online: Zika Virus, Social Media and Distrust in the Centers for Disease Control and Prevention." *Sociology of Health & Illness* 41, no. 7 (September 2019): 1270–1288. https://doi.org/10.1111/1467-9566.12924.
Leary, Warren E. "Science Takes a Lesson from Nature, Imitating Abalone and Spider Silk." *New York Times*, August 31, 1993.
Levi, Herbert S. *Spiders and Their Kin*. A Golden Nature Guide. Racine, WI: Western Pub. Co., 1968. www.jstor.org/stable/42576652.
Levine, Etan. "The Symbolism of Milk and Honey." *ETC: A Review of General Semantics* 41, no. 1 (1984): 33–37.
Lévi-Straus, Claude. *The Savage Mind*. Translated from French. London: Weidenfeld & Nicolson, 1966.
Lewis, Carol. "A New Kind of Fish Story: The Coming of Biotech Animals." *FDA Consumer* 35, no. 1 (February 2001): 14–20.
Linnaeus, C. "Araneae." In *Systema Naturae*, 1030–1037. Salvius: Holmiae, 1767.
Linzey, Andrew, and Desmond Tutu, eds. *The Global Guide to Animal Protection*. Urbana: University of Illinois Press, 2013.
Lusk, Jayson L., Mustafa Jamal, Lauren Kurlander, Maud Roucan, and Lesley Taulman. "A Meta-Analysis of Genetically Modified Food Valuation Studies." *Journal of Agricultural and Resource Economics* 30, no. 1 (2005): 28–44. https://doi.org/10.22004/AG.ECON.30782.
McFadden, Jonathan R., and Wallace E. Huffman. "Willingness-to-Pay for Natural, Organic, and Conventional Foods: The Effects of Information and Meaningful Labels." *Food Policy* 68 (April 2017): 214–232. https://doi.org/10.1016/j.foodpol.2017.02.007.
McGlotten, Shaka. "Streaking." *Drama Review* 63, no. 4 (2019): 152–171. www.muse.jhu.edu/article/740509.
Mehrabi, Tara. "Making Death Matter: A Feminist Technoscience Study of Alzheimer's Sciences in the Laboratory." PhD diss., Linköping University, 2016. https://doi.org/10.3384/diss.diva-132635.
Mendelson, Anne. *Milk: The Surprising Story of Milk through the Ages: With 120 Adventurous Recipes That Explore the Riches of Our First Food*. New York: Alfred A. Knopf, 2008.
Mirowski, Philip. *Science-Mart: Privatizing American Science*. Cambridge, MA: Harvard University Press, 2011.
Moore, Grace, Michael Tessler, Seth W. Cunningham, Julio Betancourt, and Robert Harbert. "Paleo-metagenomics of North American Fossil Packrat Middens: Past Biodiversity Revealed by Ancient DNA." *Ecology and Evolution* 10, no. 5 (March 2020): 2530–2544. https://doi.org/10.1002/ece3.6082.

Moore, Lisa Jean. *Catch and Release: The Enduring yet Vulnerable Horseshoe Crab.* New York: New York University Press, 2017.

———. "Incongruent Bodies: Teaching While Leaking." *Feminist Teacher* 17, no. 2 (2007): 95–106.

———. *Sperm Counts: Overcome by Man's Most Precious Fluid.* New York: New York University Press, 2007.

Moore, Lisa Jean, and Mary Kosut. *Buzz: Urban Beekeeping and the Power of the Bee.* New York: New York University Press, 2013.

Moore, Lisa Jean, and Matthew Allen Schmidt. "On the Construction of Male Differences: Marketing Variations in Technosemen." *Men and Masculinities* 1, no. 4 (April 1999): 331–351. https://doi.org/10.1177/1097184X99001004001.

Moore, Lisa Jean, and Rhoda Wilkie. "Introduction to *The Silent Majority*: Invertebrates in Human-Animal Studies." *Society and Animals* 27 (2019): 1–3.

Morgan, Jennifer L. *Laboring Women: Reproduction and Gender in New World Slavery.* Early American Studies. Philadelphia: University of Pennsylvania Press, 2004.

Morris, Patricia. *Blue Juice: Euthanasia in Veterinary Medicine.* Animals, Culture, and Society. Philadelphia: Temple University Press, 2012.

Murugan, Perumal, and N. Kalyan Raman. *The Story of a Goat.* First Grove Atlantic paperback ed. New York: Black Cat, 2019.

Museum of Modern of Art. *Items: Is Fashion Modern?* October 1, 2017. www.moma.org.

Nagel, Thomas. "What Is It Like to Be a Bat?" *Philosophical Review* 83, no. 4 (October 1974): 435. https://doi.org/10.2307/2183914.

Nagy, Kelsi, and Phillip David Johnson, eds. *Trash Animals: How We Live with Nature's Filthy, Feral, Invasive and Unwanted Species.* Minneapolis: University of Minnesota Press, 2013.

Nelson, Alondra. *Social Life of DNA.* Boston: Beacon, 2016.

Norman, Donald A. *The Design of Everyday Things.* Rev. and expanded ed. New York: Basic Books, 2013.

Noske, Barbara. *Beyond Boundaries: Humans and Animals.* Buffalo, NY: Black Rose Books, 1997.

Nowak, Claire. "11 Monogamous Animals That Stay Together All of Their Lives." *Reader's Digest*, December 6, 2019. www.rd.com.

Ogden, Thomas H. "Fear of Breakdown and the Unlived Life." *International Journal of Psychoanalysis* 95, no. 2 (April 2014): 205–223. https://doi.org/10.1111/1745-8315.12148.

Ombelet, W., and J. Van Robays. "Artificial Insemination History: Hurdles and Milestones." *Facts, Views and Visions in OBGYN* 7, no. 2 (2015): 137–143.

Ouellette, Jennifer/ "Along Came a Spider: The Wonders of Spider Silk." *Scientific American*, March 7, 2012. https://blogs.scientificamerican.com.

Öz, Bülent, Fahri Unsal, and Hormoz Movassaghi. "Consumer Attitudes toward Genetically Modified Food in the United States: Are Millennials Different?" *Journal of Transnational Management* 23, no. 1 (January 2, 2018): 3–21. https://doi.org/10.1080/15475778.2017.1373316.

Perlo, Katherine. "Marxism and the Underdog." *Society & Animals* 10, no. 3 (2002): 303–318, 306. https://doi.org/10.1163/156853002320770092.

Phillips, Mary T. "Savages, Drunks, and Lab Animals: The Researcher's Perception of Pain." *Society & Animals* 1, no. 1 (1993): 61–81. https://doi.org/10.1163/156853093X00154.

Pollock, Anne. *Synthesizing Hope: Matter, Knowledge, and Place in South African Drug Discovery*. Chicago: University of Chicago Press, 2019.

Punter, D. "Arachnographia: On Spidery Writing." *Storyworlds* 9, no. 1 (2017): 143–157. https://doi.org/10.5250/storyworlds.9.1-2.0143.

Putzier, Konrad. "Well-Employed in Pandemic Times: Landscaping Goats." *Wall Street Journal*, December 15, 2020. www.wsj.com.

Quammen, David. "The Virus, the Bats and Us." *New York Times*, December 11, 2020. www.nytimes.com.

Radbill, Samuel. "The Role of Animals in Infant Feeding." In *American Folk Medicine: A Symposium*, edited by Wayland Hand, 21–31. Berkley: University of California Press, 1976.

Riesman, David, Nathan Glazer, and Reuel Denney. *The Lonely Crowd: A Study of the Changing American Character*. Veritas paperback ed. New Haven, CT: Yale University Press, 2020.

Roos, Anna Marie Eleanor, ed. *The Correspondence of Dr. Martin Lister (1639–1712)*. History of Science and Medicine Library, 48: 202. Boston: Brill, 2015.

———. *Martin Lister and His Remarkable Daughters: The Art of Science in the Seventeenth Century*. Oxford: Bodleian Library, University of Oxford, 2019.

Roosth, Sophia. *Synthetic: How Life Got Made*. Chicago: University of Chicago Press, 2017.

Rosenberg, Martha. "Despite Its Celebrity Backers, Charity Heifer International Does More Harm than Good." *AlterNet*, November 3, 2017. www.alternet.org.

———. "Live Animal Charities Are Ill Conceived; Increase Suffering." USC Annenberg. *Center for Health Journalism* (blog), January 15, 2013. www.centerforhealthjournalism.org.

Roy, Deboleena. "Asking Different Questions: Feminist Practices for the Natural Sciences." *Hypatia* 23, no. 4 (December 2008): 134–157. https://doi.org/10.1111/j.1527-2001.2008.tb01437.x.

Runge, Kristin K., Dominique Brossard, Dietram A. Scheufele, Kathleen M. Rose, and Brita J. Larson. "Attitudes about Food and Food-Related Biotechnology." *Public Opinion Quarterly* 81, no. 2 (2017): 577–596. https://doi.org/10.1093/poq/nfw038.

Rutherford, Adam. *Creation: How Science Is Reinventing Life Itself*. New York: Current, 2013, 150.

———. "Synthetic Biology." In *What the Future Looks Like*, edited by Jim Al-Khalili, 75–84. New York: The Experiment, 2017.

Santhanam, Laura. "Gun Deaths Started to Rise After More than a Decade of Being Stable." *PBS News Hour* (blog), October 9, 2019. www.pbs.org.

Saravanan, D. "Spider Silk: Structure, Properties and Spinning." *Journal of Textile and Apparel, Technology and Management* 5, no. 1 (winter 2006).

Schmitt, W. J., and R. M. Muri. "Neurobiologie Der Spinnenphobie. Schweizer." *Archiv Fur Neurologie Und Psychiatrie* 160, no. 8 (2009): 352–355.

Schuller, Kyla. *The Biopolitics of Feeling: Race, Sex, and Science in the Nineteenth Century*. Anima. Durham, NC: Duke University Press, 2018.

Scott, Sydney E., Yoel Inbar, and Paul Rozin. "Evidence for Absolute Moral Opposition to Genetically Modified Food in the United States." *Perspectives on Psychological Science* 11, no. 3 (May 2016): 315–324. https://doi.org/10.1177/1745691615621275.

Scott, Sydney E., Yoel Inbar, Christopher D. Wirz, Dominique Brossard, and Paul Rozin. "An Overview of Attitudes toward Genetically Engineered Food." *Annual Review of Nutrition* 38, no. 1 (August 21, 2018): 459–479. https://doi.org/10.1146/annurev-nutr-071715-051223.

Shapin, Steven. *The Scientific Life: A Moral History of a Late Modern Vocation*. Chicago: University of Chicago Press, 2010.

Shapiro, Beth Alison, *How to Clone a Mammoth: The Science of De-extinction*. Princeton, NJ: Princeton University Press, 2016.

Sharp, Lesley A. "Animal Research Unbound: The Messiness of the Moral and the Ethnographer's Dilemma." *History and Philosophy of the Life Sciences* 43, no. 2 (June 2021): 76. https://doi.org/10.1007/s40656-021-00426-2.

Simoons, Frederick, and John Baldwin. "Breast-Feeding of Animals by Women: Its Socio-Cultural Context and Geographical Occurrence." *Anthropos* 77, no. 3/4 (1982): 421–448.

Singh, D. R., Sushila Kaul, and N. Sivaramane. "Migratory Sheep and Goat Production System: The Mainstay of Tribal Hill Economy in Himachal Pradesh," 2006. https://doi.org/10.22004/AG.ECON.57771.

Soth, Amelia. "The Tangled History of Weaving with Spider Silk." *JSTOR Daily*, November 15, 2018. https://daily.jstor.org.

Star, Susan Leigh, and James R. Griesemer. "Institutional Ecology, `Translations' and Boundary Objects: Amateurs and Professionals in Berkeley's Museum of Vertebrate Zoology, 1907–1939." *Social Studies of Science* 19, no. 3 (August 1989): 387–420. https://doi.org/10.1177/030631289019003001.

Starbard, Ann. *The Dairy Goat Handbook: For Backyard, Homestead, and Small Farm*. Minneapolis: Voyageur Press, 2015.

Stengers, Isabelle. "Autonomy and the Intrusion of Gaia." *South Atlantic Quarterly* 116, no. 2 (April 1, 2017): 381–400. https://doi.org/10.1215/00382876-3829467.

———. *Cosmopolitics*. Posthumanities 9–10. Minneapolis: University of Minnesota Press, 2010.

Stewart, Kathleen. "On the Politics of Cultural Theory: A Case for 'Contaminated' Cultural Critique." *Social Research* 58, no. 2 (1991): 395–412.

Strauss, Anselm, and Juliet Corbin. *Basics of Qualitative Research*. Newbury Park, CA: SAGE Publications, 1990.

Sunder Rajan, Kaushik, ed. *Lively Capital: Biotechnologies, Ethics, and Governance in Global Markets*. Experimental Futures. Durham, NC: Duke University Press, 2012.

Takamoto, Iwao, and Charles Nichols, directors. *Charlotte's Web*. Animated film. Paramount Pictures and Hanna Barbera Productions, 1973.

Taussig, Karen-Sue. "Bovine Abominations: Genetic Culture and Politics in the Netherlands." *Cultural Anthropology* 19, no. 3 (2004): 305–336. https://doi.org/10.1525/can.2004.19.3.305.

Taylor, Nik. "'Never an It': Intersubjectivity and the Creation of Animal Personhood in Animal Shelters." *Qualitative Sociological Review* 3, no. 1 (2007): 59–73.

Taylor, Sunaura. *Beasts of Burden: Animal and Disability Liberation*. New York: New Press, 2017.

Tonnelat, Stéphane. "'Out of Frame': The (in)Visible Life of Urban Interstices—A Case Study in Charenton-Le-Pont, Paris, France." *Ethnography* 9, no. 3 (September 2008): 291–324. https://doi.org/10.1177/1466138108094973.

Tsing, Anna. *The Mushroom at the End of the World: On the Possibility of Life in Capitalist Ruins*. Princeton, NJ: Princeton University Press, 2015.

US Department of Agriculture. "Sheep and Goats." Census of Sheep and Goats. USDA, February 2019. www.nass.usda.gov.

US Department of Agriculture, and Animal and Plant Health Inspection Service. "Animal Welfare Act and Animal Welfare Regulations." 1966. www.nal.usda.gov.

US Food and Drug Administration. "Animals with Intentional Genomic Alterations." *Animal and Veterinary*, December 14, 2020. www.fda.gov.

———. "Guidance for Industry Regulation of Intentionally Altered Genomic DNA in Animals." Guidance No. 187. January 2017. www.fda.gov.

Van Haren, Juliette. "Why We Need Alternatives to Petro-Chemical-Based Materials." *New School—Parsons Health Materials Lab* (blog), November 2, 2018. https://healthymaterialslab.org.

Wang, Ming, Zhaolin Sun, Tian Yu, Fangrong Ding, Lijng Li, Xi Wang, and Fu Mingbo. "Large-Scale Production of Recombinant Human Lactoferrin from High-Expression, Marker-Free Transgenic Cloned Cows." *Nature: Scientific Reports* 7, no. 10,733 (2017). www.nature.com.

Weber, Larry. *Web Watching: A Guide to Webs & the Spiders That Make Them*. Stone Ridge Press, 2018.

Weinbaum, Alys Eve. *The Afterlife of Reproductive Slavery: Biocapitalism and Black Feminism's Philosophy of History*. Durham, NC: Duke University Press, 2019.

Whatmore, Sarah. "Materialist Returns: Practising Cultural Geography in and for a More-than-Human World." *Cultural Geographies* 13, no. 4 (October 2006): 600–609. https://doi.org/10.1191/1474474006cgj377oa.

White, E. B., and Garth Williams. *Charlotte's Web*. New York: HarperCollins, 1952.

Whitley, Cameron Thomas, Linda Kalof, and Tim Flach. "Using Animal Portraiture to Activate Emotional Affect." *Environment and Behavior*, June 9, 2020: 837–863. https://doi.org/10.1177/0013916520928429.

Wilkie, Rhoda. *Livestock/Deadstock: Working with Farm Animals from Birth to Slaughter*. Philadelphia: Temple University Press, 2010.

Wilkie, Rhoda, Lisa Jean Moore, and Claire Molloy. "How Prevalent Are Invertebrates in Human-Animal Scholarship? Scoping Study of Anthrozoos and Society & Animals." *Society and Animals* 27 (2019): 4–29. https://doi.org/10.1163/15685306-00001902.

Williams, Raymond. *The Country and the City*. London: Chatto & Windus Ltd., 1973.

———. "The Technology and the Society (1974)." In *Raymond Williams on Culture and Society: Essential Writings*, edited by Jim McGuigan, 139–160. London: SAGE Publications Ltd., 2014. https://doi.org/10.4135/9781473914766.n9

Winnicott, Clare, Ray Shepherd, and Madeleine Davis, eds. "Fear of Breakdown." In *D. W. Winnicott: Psycho-Analytic Explorations*, 87–95. Cambridge, MA: Harvard University Press, 1989.

Winnicott, Donald W. *The Child, the Family, and the Outside World*. A Merloyd Lawrence Book. Cambridge, MA: Perseus 1987.

Yanagisako, Sylvia Junko, Carol Lowery Delaney, and American Anthropological Association, eds. *Naturalizing Power: Essays in Feminist Cultural Analysis*. New York: Routledge, 1995.

Yusoff, Kathryn. "Aesthetics of Loss: Biodiversity, Banal Violence and Biotic Subjects." *Transactions* 37, no. 4 (2011): 578–592.

Zemlin, J. C. "A Study of the Mechanical Behavior of Spider Silks." Technical Report. Waltham, MA: US Army Natick Laboratories, September 1968.

Zhang, Shichang, Dakota Piorkowski, Wan-Rou Lin, Yi-Ru Lee, Chen-Pan Liao, Pi-Han Wang, and I-Min Tso. "Nitrogen Inaccessibility Protects Spider Silk from Bacterial Growth." *Journal of Experimental Biology* 222, no. 20 (October 15, 2019): jeb214981. https://doi.org/10.1242/jeb.214981.

INDEX

alfalfa, 19, 82, 84, 93, 124, 140, 175n42, 187n17
alienation, 12, 94, 102, 107, 124–125
American Dairy Goat Association (ADGA), 67
animals as capital, 1–2, 9, 29, 65, 126
animal studies. *See* critical animal studies
anthropology, 2–3, 5, 10, 17, 25, 74, 100, 102, 106, 119, 125, 173n6
anthropomorphism, 24, 50, 64, 66
arachnology, 14, 26, 37–38, 176n49
arachnophobia, 26, 33–35, 40, 50, 137, 163, 188n25
assisted reproduction. *See* reproduction: technologies
Atwood, Margaret, 29, 162. *See also* dystopia

banana spiders. *See* golden orb weaver spiders
Beebe Lab. *See* Eighteen B
bees, 14, 25, 29, 32, 34, 41, 43, 50, 165
Bell, Brianne, 94–96, 108, 112, 114–116, 140, 183n14, 183n16, 184n22
bioengineered organism. *See* systems
biomimetics (biomimicry), 28, 119, 176n49
biotechnology, 2, 11, 14, 26, 76, 88, 91, 93, 162, 181n28
bodies, 3, 74, 89, 134–139, 158, 179–180n8; of animals, 7, 38–41, 45, 59, 63, 79, 89, 96–97, 101, 111; embodiment (and disembodiment), xii, 3–4, 15, 25, 34, 101, 105, 157; of women, 5, 7–8, 63, 72, 80, 111, 143, 158

Bolt Threads, 57, 135, 144–148, 176n49, 186n7, 188n25. *See also* Eighteen B
breeding. *See* reproduction: management of
Breslauer, David, 57–58, 145–147, 176n49
b-silk, 145, 148, 188n25

capitalism, x, 2, 8–9, 12, 26, 101, 107, 110, 120–126, 131–133, 136, 139, 143, 150–151, 156, 159, 161–162
Capra aegagrus hircus. *See* Saanen goats
Charlotte's Web, 31, 36, 47–48, 176n4
cloning, 2, 12, 76, 96
COVID-19 pandemic, x, 129, 136, 150, 154, 166–167
critical animal studies, 2–3, 17, 26–27, 32, 47, 63, 74
culling (of goats), 27, 64, 79, 85, 87, 93–97, 127, 161

Darwin, Charles, 101, 103, 105–108, 122–124
DNA, ix, xi, xvii, 3–4, 9–10, 25, 75, 87, 89, 104–105, 119, 123, 132, 137, 160, 181n34; of spiders, ix, xvi, 59, 76, 188n25
Dolly the Sheep. *See* cloning
dystopia, x, 29. *See also* Atwood, Margaret

Edwards, G. B., 49–51, 54, 178n38
Eighteen B, 148–149, 188n25. *See also* Bolt Threads
entomology, 2, 14, 25, 27, 43–44, 49
Enviropig, 9
Escherichia coli (*E. coli*), 9, 18, 56, 83, 91, 93, 104–105, 124, 140, 159–160, 179n48, 186–187n15, 187n17

ethnography, xi–xii, 2–4, 14–15, 54, 86, 102, 126, 157, 164
euthanasia. *See* culling
exploitation, x, 3, 8, 55, 75, 78–79, 86, 124, 136, 147, 162–163, 165, 168
extinction, 34, 119–120, 160–162

false consciousness (and consciousness-raising), 4, 17, 48, 63, 73, 106, 145, 147, 153–154, 164
feminism, x, xiii, 2–5, 7, 15, 26, 34, 50, 103, 150, 156
Food and Drug Administration, United States (FDA), xvii, 27, 84, 88–90, 97, 132, 141–142, 153, 167, 182n45, 187n19
Franklin, Sarah. *See* cloning

Gainesville, Florida, 14, 25–27, 34, 36, 42–49, 116, 162–163, 176n49
Gasinby, Danielle, 136, 165
gender roles, x, xii, 12, 41, 157. *See also* normativity
gender studies, xiii, 2–3
golden orb weaver spiders, 26–27, 32, 34–35, 38–48, 50–57, 59, 116, 132–133, 147, 163, 176n49, 176n4, chap 1, 177n13, 178n38
grounded theory, 14, 175–176n38, 188n28. *See also* reflexivity

Harris, Thomas, 19
Hayashi, Cheryl, 25, 39–40, 52, 54–55, 57–58, 145, 177n13
heteronormativity. *See under* normativity
heterosexuality, xii, 17, 41–42, 49–50, 99
honeybees. *See* bees
horseshoe crabs, 1–2, 4, 14, 25, 32, 34, 41, 43, 50, 106, 136, 165, 176n6

innovation, ix, xi, 1, 7, 9, 12, 26, 28, 59, 91, 93, 101, 121–123, 129, 131–132, 135–136, 139–140, 142, 146, 149–150, 159, 162, 176n49

insemination. *See* reproduction; semen; sperm
intellectual property, 28, 78, 114, 121–122, 185n39
invertebrates, 1, 3, 26, 32, 34–35, 50, 54, 163
Iverson, Christian, 121, 140–141, 144

Jones, Andrew, 22, 85, 87–88
Jones, Justin, 40, 55, 61, 79–85, 88–89, 91–93, 96–97, 110–111, 113, 121, 134–135, 140–141, 143, 145, 147, 159, 175n42, 179n48, 181n38, 187n17

Kevlar, xvii, 55, 130, 146, 148, 185n1
Kuntner, Matjaz, 38–40, 176n49

labor, xiv, 2, 8, 29, 52, 62, 81, 87, 110, 117, 125–126, 151
lactation, ix, 1, 5, 9, 32, 61–62, 66–69, 71–72, 75, 79, 83, 97, 101–102, 127, 150, 157–158, 179nn1–6, 180n21, 181n35. *See also* milking
Lewis, Randolph "Randy," 14, 18, 22, 28, 39, 52, 55–56, 58, 76–78, 81–84, 92–93, 97, 113, 121, 142–145, 153–154, 164–165
Logan, Utah, 14–18, 22, 26, 162

machines, 68, 108–109, 111–117, 125, 132, 168; animals as, 2, 24–25, 75–76, 90, 101, 107, 114, 126, 137, 139
major spider ampullate protein (MaSp), 40, 80, 88, 109, 112, 124, 140, 183n14
Marxism, 124–126, 151, 159
maternalism. *See* motherhood
McCartney, Stella, 147
milking (of goats), 4, 22, 24, 27, 57, 59, 62–63, 67–70, 74–77, 79–80, 85–87, 90–91, 94–95, 97, 101, 108, 111, 125, 132, 153, 156, 158, 167–168. *See also* lactation
motherhood, x–xiii, 9, 26, 45, 61–63, 71–74, 82, 137–138, 150, 155, 157, 162–163, 165
Murugan, Perumal, 78–79

naturalness, xi–xiii, 6–7, 9–10, 18, 27, 41, 51, 99–101, 106–107, 110–111, 124, 126, 158, 164–165–168, 184n21; the natural world, x, xvii, 6, 12, 42, 119, 123, 168
neoliberalism. *See* capitalism
New York City, xii, 14–18, 25–26, 39, 103, 147, 162
Nexia Biotechnologies, 76, 146
normativity, xii, 6, 17, 66; heteronormativity, 12, 41–42, 100–101, 157, 161

obsolescence, 1, 26, 93, 97, 160, 162

patents. *See* intellectual property
psychoanalysis, 135, 137, 186n8, 186n11
purification, 19, 26, 28, 81, 91, 93, 97, 100–102, 107–118, 183n14
purity, x, xiv, 7, 9–11, 99–107, 110–112, 116, 122, 126, 135, 154, 161, 184n20

queerness, x–xiii, 16–18, 25–26, 49, 99–100, 103, 155, 157, 163, 182n3

race and racialization, 5, 8–9, 41, 71, 103, 110, 157, 174n19, 184n18, 185n46
Reeves, Lawrence "Lary," 43–48, 177n12
reflexivity, 5, 13–15, 157. *See also* grounded theory
reproduction, x, xiii, 6–8, 13, 17, 26–27, 41, 61, 69, 80, 102, 111, 118, 155, 158, 161, 165, 174n19; in goats, 64–65, 75–76, 79–82, 93–94, 124, 127, 132, 153, 157, 163, 180n21; in golden orb weaver spiders, 36, 44–47, 49–52; as labor, 2, 8, 78–79, 111; management of, 7–8, 10, 27, 64, 67, 75, 78–82, 92, 94, 97, 101, 116, 129, 132, 163–164; queer, xi, xiii, 24, 163; technologies, xi–xii, 6, 11–12, 18, 51–52, 61, 63, 90, 99–101, 119, 158, 165–166, 168, 182nn1–2, 182–183n4
risk, xii, 11–12, 89, 129, 131, 134–135, 167
Rudenko, Larisa, 88–90

Saanen goats, 66–67, 108
scalability, 1, 19, 37, 90, 122, 127, 141, 147, 160
semen, x–xii, 11, 52, 99–100, 158; technosemen, xi–xii, 11, 100–101, 121. *See also* sperm
silking, 53–54, 57–59, 117, 132, 145. *See also* spider silk, production of
silkworms, 19, 93, 124, 135, 140, 147, 160, 175n42, 187n17
sociology, xi–xiv, 2–3, 12–14, 16, 27, 88, 99–101, 107, 110, 122, 126, 132–133, 150, 164–165
South Farm, 14, 22, 24–25, 67–69, 74, 79, 85–86, 90, 108, 111, 167
speciesism, 33–34, 63, 72, 74, 165
sperm, xi–xii, 18, 27, 29, 51–52, 90, 100–101; sperm donation, xi–xii, 11, 18, 29, 63, 90, 99–100, 155, 158, 165. *See also* semen
spider goats, profitability of, 1–2, 28–29, 91–93, 101–102, 120–122, 126–127, 129, 158–159, 186–187n15
Spider-Man, x, 18–20, 32, 34, 36, 44, 49
Spider Silk Laboratory, 13–14, 18–23, 28, 40, 52, 55, 76, 79, 81, 83–84, 88, 91, 94, 102, 108–116, 122–124, 132, 136, 140, 142, 144–145, 147, 158–159, 165, 179n48
spider silk, mechanical properties of, 40, 53, 55, 160
spider silk, production of (by spiders), 35–39, 177n12, 177n14. *See also* silking
spider silk, uses of, ix, xvi, 35, 37, 120, 123–124, 127, 129–151, 158, 175n42, 181n38, 187n17; in adhesives, xvii, 2, 28, 77, 134–136, 139–141; in cosmetics, 1, 26, 144–145, 147–150; in fabric or apparel, ix, 1, 26, 28, 37, 52, 55, 77, 123, 129, 131, 133, 135, 139, 144–147, 150; medical, ix, xvii, 1, 28–29, 77, 119, 123, 127, 129, 131–132, 134–136, 139–144, 150, 159; military, ix, xvii, 1, 26, 28–29, 35, 52–55, 76–77, 90, 119, 127, 129, 131–132, 139, 146–147, 159

Stella McCartney. *See* McCartney, Stella
Story of a Goat, The. *See* Murugan, Perumal
synthetic biology, 12, 28, 102, 104, 108, 118–123, 144–145, 150, 158–159
systems, 18–19, 56, 63, 81, 83, 93, 104, 119, 123–127, 140, 158–161, 166, 175n42; goat system, 18–19, 28, 76, 81, 83, 91, 93, 99–127, 140, 158–159, 175n42, 185n39

Thorton, Amber, 22, 80, 85–88, 94
Trichonephila clavipes. *See* golden orb weaver spiders

United States Animal Welfare Act of 1966, 54, 88
United States Department of Agriculture (USDA), 24, 27, 65, 84, 88, 97, 132, 153, 184n18
United States military, ix, xvii, 28–29, 35, 52–55, 77, 90, 126, 130, 132, 139, 159
University of Wyoming (UW), 76, 78, 82–83, 97, 110, 132
Utah Science Technology and Research Initiative (USTAR), 14, 18, 120
Utah State University (USU), 14, 22, 78, 87, 96–97, 120–122, 132, 140
Utah State University Animal Science Farm. *See* South Farm

vaccines, x, 11–12, 136, 166–167

White, John, 67, 71–72
Winnicott, Donald, 137–138, 186n8–10

yeast, 119, 146, 148, 160, 187n17

Zhang, Xiaoli, 19
zoonosis. *See* COVID-19 pandemic

ABOUT THE AUTHOR

LISA JEAN MOORE is a medical sociologist and SUNY Distinguished Professor of Sociology and Gender Studies in the School of Natural and Social Sciences at Purchase College, SUNY. Her books include an ethnography of honeybees, *Buzz: Urban Beekeeping and the Power of the Bee* (New York University Press), coauthored with Mary Kosut. *Catch and Release: The Enduring, yet Vulnerable, Horseshoe Crab* (New York University Press) examines the interspecies relationships between humans and *Limulus polyphemus* (Atlantic horseshoe crabs). She lives in Brooklyn, New York, with her family.

Lightning Source UK Ltd.
Milton Keynes UK
UKHW031847010622
403847UK00002B/321